United States Nuclear Regulatory Commission

Protecting People and the Environment

NUREG/CR-7164
ORNL/TM-2012/518

I0493843

Cross Section Generation Guidelines for TRACE-PARCS

Office of Nuclear Regulatory Research

AVAILABILITY OF REFERENCE MATERIALS
IN NRC PUBLICATIONS

NRC Reference Material

As of November 1999, you may electronically access NUREG-series publications and other NRC records at NRC's Public Electronic Reading Room at http://www.nrc.gov/reading-rm.html. Publicly released records include, to name a few, NUREG-series publications; *Federal Register* notices; applicant, licensee, and vendor documents and correspondence; NRC correspondence and internal memoranda; bulletins and information notices; inspection and investigative reports; licensee event reports; and Commission papers and their attachments.

NRC publications in the NUREG series, NRC regulations, and Title 10, "Energy," in the *Code of Federal Regulations* may also be purchased from one of these two sources.
1. The Superintendent of Documents
   U.S. Government Printing Office Mail Stop SSOP
   Washington, DC 20402–0001
   Internet: bookstore.gpo.gov
   Telephone: 202-512-1800
   Fax: 202-512-2250
2. The National Technical Information Service
   Springfield, VA 22161–0002
   www.ntis.gov
   1–800–553–6847 or, locally, 703–605–6000

A single copy of each NRC draft report for comment is available free, to the extent of supply, upon written request as follows:
Address: U.S. Nuclear Regulatory Commission
         Office of Administration
         Publications Branch
         Washington, DC 20555-0001
E-mail: DISTRIBUTION.RESOURCE@NRC.GOV
Facsimile: 301–415–2289

Some publications in the NUREG series that are posted at NRC's Web site address http://www.nrc.gov/reading-rm/doc-collections/nuregs are updated periodically and may differ from the last printed version. Although references to material found on a Web site bear the date the material was accessed, the material available on the date cited may subsequently be removed from the site.

Non-NRC Reference Material

Documents available from public and special technical libraries include all open literature items, such as books, journal articles, transactions, *Federal Register* notices, Federal and State legislation, and congressional reports. Such documents as theses, dissertations, foreign reports and translations, and non-NRC conference proceedings may be purchased from their sponsoring organization.

Copies of industry codes and standards used in a substantive manner in the NRC regulatory process are maintained at—
         The NRC Technical Library
         Two White Flint North
         11545 Rockville Pike
         Rockville, MD 20852–2738

These standards are available in the library for reference use by the public. Codes and standards are usually copyrighted and may be purchased from the originating organization or, if they are American National Standards, from—
         American National Standards Institute
         11 West 42nd Street
         New York, NY 10036–8002
         www.ansi.org
         212–642–4900

Legally binding regulatory requirements are stated only in laws; NRC regulations; licenses, including technical specifications; or orders, not in NUREG-series publications. The views expressed in contractor-prepared publications in this series are not necessarily those of the NRC.

The NUREG series comprises (1) technical and administrative reports and books prepared by the staff (NUREG–XXXX) or agency contractors (NUREG/CR–XXXX), (2) proceedings of conferences (NUREG/CP–XXXX), (3) reports resulting from international agreements (NUREG/IA–XXXX), (4) brochures (NUREG/BR–XXXX), and (5) compilations of legal decisions and orders of the Commission and Atomic and Safety Licensing Boards and of Directors' decisions under Section 2.206 of NRC's regulations (NUREG–0750).

U.S.NRC

United States Nuclear Regulatory Commission

Protecting People and the Environment

NUREG/CR-7164
ORNL/TM-2012/518

Cross Section Generation Guidelines for TRACE-PARCS

Manuscript Completed: November 2012
Date Published: June 2013

Prepared by:
D. Wang, B. J. Ade, and A. M. Ward

Oak Ridge National Laboratory
P. O. Box 2008
Oak Ridge, TN 37831-6010

C. G. Thurston, NRC Project Manager

NRC Job V6233

Office of Nuclear Regulatory Research

ABSTRACT

This report documents a comprehensive comparison of cross sections calculated using different methodologies and codes, including CASMO, HELIOS, and TRITON XS. The conclusions from this study have resulted in this guidance document on how to choose cross section histories and branches for boiling water reactor (BWR) analysis, and the methodology to collapse the fine-energy and -space fluxes calculated by the detailed lattice calculation. The guidance herein is applicable to all BWR designs.

For BWR steady-state and transient analysis, the PARCS code uses two-energy-group cross sections for each computational node in the 3-dimensional grid. The PARCS cross sections are tabulated as a function of four instantaneous state variables: (1) control rod insertion, (2) fuel temperature, (3) coolant density, and (4) soluble poison concentration. The cross section values also depend on the isotopic mixture (i.e., concentration of ^{235}U, ^{239}Pu ...), which is characterized as a function of control rod and moderator density history variables.

In a typical calculation, the fuel temperature, moderator density, and soluble boron concentration are calculated by the TRACE code for a coupled TRACE/PARCS analysis. The instantaneous control rod insertion is provided by the user in the input deck. The historic control rod and moderator density values are provided by a steady-state core-follow simulator, which has followed the core operation since the initial loading up to the time of the transient to be calculated. All these parameters are taken into account to estimate the instantaneous cross section based on the tabulated values. This report documents the expected error in evaluating the instantaneous cross section as a function of the data table structure.

As a result of this study, guidelines for BWR cross section generation have been generated. The recommendations are the use of four instantaneous moderator density values at 0, 40, 70, and 90% void fraction at three different fuel temperatures of 500, 950, and 1500 K. For the history effect, three moderator density values at 0, 40, and 70% at a single 950 K fuel temperature provide sufficiently accurate results. For all cases, the user must generate these branches for controlled and uncontrolled bundles.

In addition, a coolant density branch with a moderator density of 1000 kg/m^3 is needed to accurately model cold depressurized conditions. If boron injection is modeled in a BWR transient analysis, boron branches should be included.

TABLE OF CONTENTS

Section	Page

LIST OF FIGURES

LIST OF TABLES

ACRONYMS AND ABBREVIATIONS

BC	boundary condition
BOL	beginning of life
BWR	boiling water reactor
CR	control rod insertion
DC	coolant density
DF	discontinuity function
EOL	end of life
GenPMAXS	Generation of the Purdue Macroscopic XS Set
HCR	control rod history
HDC	history of coolant density
ORNL	Oak Ridge National Laboratory
PARCS	Purdue Advanced Reactor Core Simulator
PC	poison concentration
pcm	percent mil (or 10^{-5})
RPV	reactor pressure vessel
TF	fuel temperature
TRACE	TRAC/RELAP Advanced Computational Engine
XS	cross sections

1.0   INTRODUCTION

In recent years, there has been increased application of TRACE/PARCS for light water reactor transient safety analysis, in particular for boiling water reactor (BWR) stability analysis. TRACE is a thermal-hydraulics system analysis code. PARCS solves time-dependent spatial kinetics equations based on the two-group diffusion nodal method.

A BWR core usually consists of hundreds of fuel assemblies. The coolant density (DC) changes significantly from the core inlet to the outlet. The coolant at the core inlet is subcooled and has a density of about 0.76 g/cm^3. The DC decreases from inlet to outlet as a result of boiling as the coolant flows through the core. At the core exit, DC decreases to approximately 0.2 g/cm^3 on average with a void faction of about 75%. Therefore, for a BWR core operated with a wide distribution of DCs, historical effects of neutron spectrum on cross sections (XS) due to DC should be considered in generating XS. In addition, the effects of control rod history (HCR) on XS should be considered.

A typical BWR fuel assembly is made up of several axial lattice segments (or zones) with different enrichments and burnable poisons. For each fuel segment, two-group XS are generated using a lattice physics code (e.g., TRITON, CASMO, or HELIOS). GenPMAXS is employed to process the two-group XS and kinetics data generated with the lattice code and store them in a PMAXS file, which can be read by PARCS. PMAXS provides all of the data necessary to perform core simulations for steady-state and transient applications, consisting of principal macroscopic XS, microscopic XS of Xe/Sm, group-wise form functions, and the kinetics data.

The two-group XS for each computational node in the PARCS 3-D core model are typically parameterized as a function of five state variables: control rod insertion (CR), fuel temperature (TF), DC, and soluble poison concentration (PC), as well as control rod and moderator density history variables. Thermal-hydraulic variables such as fuel temperature, moderator density, and soluble boron concentration, are typically calculated by TRACE.

The main objective of this study of BWR cross sections is to answer the following questions:

1. How many histories and branches in the XS set are sufficient to have converged results for BWR analysis?

2. How sensitive are BWR analysis results to XS generated with different lattice codes (TRITON, CASMO, or HELIOS)?

A Peach Bottom fuel type with the 7×7 lattice was chosen for this study. Two-group XS and kinetics parameters for this fuel were generated with CASMO, HELIOS, and TRITON, respectively. Each of the XS sets has a very fine structure of histories and branches, which consists of 11 uncontrolled DC histories (HDCs) and 11 controlled HDCs; each history has 11 uncontrolled DC branches and 11 controlled DC branches.

Three BWR models employed to conduct this study are (1) Single-CHAN model, (2) Oskarshamn plant model, and (3) Ringhals plant model.

The results of the comprehensive comparison performed in this study on the three XS sets generated with CASMO, HELIOS, and TRITON show that, at beginning of life (BOL), TRITON slightly over predicts Σ_a^2 and $\nu\Sigma_f^2$ for high voids compared with CAMSO and HELIOS; overall, it has about 300 pcm of over prediction of Kinf. At the end of life (EOL) of each coolant history, TRITON over predicts Kinf by 200—1500 pcm, and the over prediction increases as void fraction increases. In addition, TRITON slightly under predicts Σ_{tr}^1 by 2–8%. However, it should be noted that the discrepancies found with TRITON do not necessarily imply that TRITON predicts these XS poorly, because the predictability of lattice codes, including CASMO and HELIOS, for coolant with high void fraction (> 80%) needs further investigation. The TRITON calculations used SCALE version 6.1; Oak Ridge National Laboratory (ORNL) is in the process of performing bias assessments with Monte Carlo codes to improve the accuracy of future releases.

Both Σ_a^1 and $\nu\Sigma_f^1$ curves show more nonlinearity at high void fraction (>70%) than at low void fraction. Thus the extrapolation of the XS from the 70% void may overestimate the XS for high void faction.

The sensitivity analysis performed on HDC effects shows that neither BWR steady-state nor transient analysis is highly sensitive to HDC effects (nor HCR effects). It is found that three HDCs at 0%, 40%, and 70% (or 80%) voids are good enough for most BWR applications.

The sensitivity analysis of DC instantaneous branch effects also shows that BWR steady-state results are not highly sensitive to DC branch effects. However, transients may be sensitive to the structure of DC branches, especially when the core upper region has void fractions that are outside the range of DC branches. The study has concluded that four DC branches at 0%, 40%, 70%, and 90% voids provide sufficient accuracy for BWR transient analysis. The 100% void branch may be used to replace the 90% branch if XS at 100% void can be calculated accurately, but a Monte Carlo type calculation is required to confirm the accuracy of the lattice code. For controlled DC branches, it is suggested that the same density branch structure be used.

2.0   FUEL SPECIFICATIONS FOR THIS STUDY

2.1.   SPECIFICATIONS OF PEACH BOTTOM TYPE 3C FUEL

Type-3c fuel was used from Peach Bottom Unit 2 Cycles 1 and 2 [1]. The type-3c fuel is a 7×7 fuel lattice with fuel, gap, and cladding outer diameters of 0.477, 0.526, and 0.563 in. (1.211156, 1.33604, and 1.43002 cm), respectively. The fuel pin pitch is 0.738 in. (1.87452 cm). The channel outer dimension is 5.438 in. (13.8152 cm) with a thickness of 0.08 in. (0.2032 cm) and a corner inner radius of 0.38 in. (0.9652 cm). Type 3c is the middle segment of the type-3 fuel bundle, as shown in Figure 1. Among the 49 fuel rods, 5 rods contain burnable poison Gd_2O_3. The enrichment of each fuel pin and the weight percentage of Gd_2O_3 are indicated in Figure 1. Detailed specifications for the control blade are given in Table 1. Figure 2 expands the control blade model in more detail.

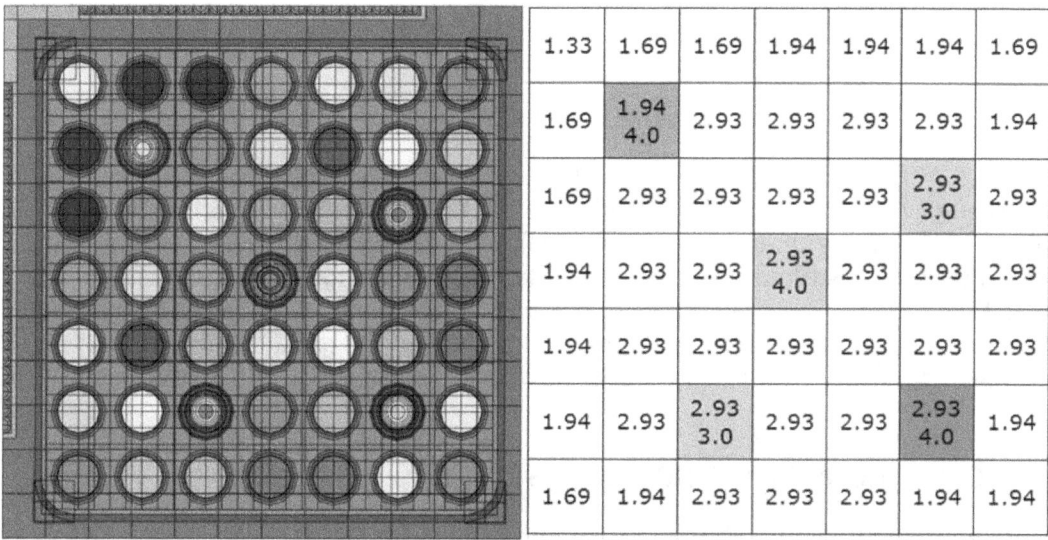

1.33	1.69	1.69	1.94	1.94	1.94	1.69
1.69	1.94 4.0	2.93	2.93	2.93	2.93	1.94
1.69	2.93	2.93	2.93	2.93	2.93 3.0	2.93
1.94	2.93	2.93	2.93 4.0	2.93	2.93	2.93
1.94	2.93	2.93	2.93	2.93	2.93	2.93
1.94	2.93	2.93 3.0	2.93	2.93	2.93 4.0	1.94
1.69	1.94	2.93	2.93	2.93	1.94	1.94

Figure 1. Type-3c fuel bundle.

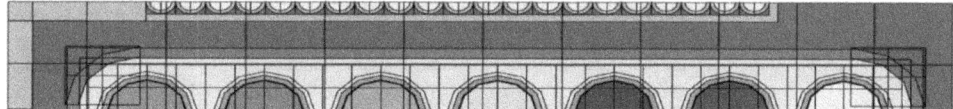

Figure 2. Control blade detail.

Table 1. Control blade design parameters

Parameter	Specification
Control material	B4C
Material density (g/cc)	1.764
Control pin tube ID (cm)	0.35052
Control pin tube ID (cm)	0.47752
Control pin tube material	SS-304 (7.94 g/cm^3)
Control blade half span (cm)	12.3825
Control blade full thickness (cm)	0.79248
Control blade tip radius (cm)	0.39624
Control blade sheath material	SS-304 (7.94 g/cm^3)
Control blade sheath thickness (cm)	1.23698
Central structure wing length (cm)	1.98501
Total number of control pins	84

2.2. SPECIFICATIONS OF PMAXS STRUCTURE

Three XS sets (PMAXS) for the PB type-3c fuel were generated with CASMO, HELIOS, and TRITON, respectively. For the CASMO and HELIOS XS sets, the PMAXS files consist of 12 uncontrolled HDCs (CR = 0) and 12 controlled HDCs (CR = 1), each history having 12 DC branches. The PMAXS structure of histories and branches is shown in Figure 3. The DC ranges from 0.03591 g/cm^3 (100% void) up to 0.73808 g/ cm^3 (0% void) in 10% void spacing. The CASMO and HELIOS sets include one additional density at 1.0 g/ cm^3 that represents cold core, depressurized conditions. The three XS sets have five TF branches at 500, 750, 100, 1250, and 1500 K.

2.3. SPECIFICATIONS OF LATTICE CODES

The version of each lattice code is given in Table 2. CAMSO and HELIOS use the ENDV-VI library, and TRITON uses the ENDF-VII library.

Table 2. Code versions and Cross Section library

	Version	Cross Section Library
CASMO	4E	ENDF-VI, 70G neutron + 18G gamma
HELIOS	1.10	ENDF-VI, 47G neutron + 17G gamma
TRITON	SCALE 6.1	ENDF-VII, 49G neutron

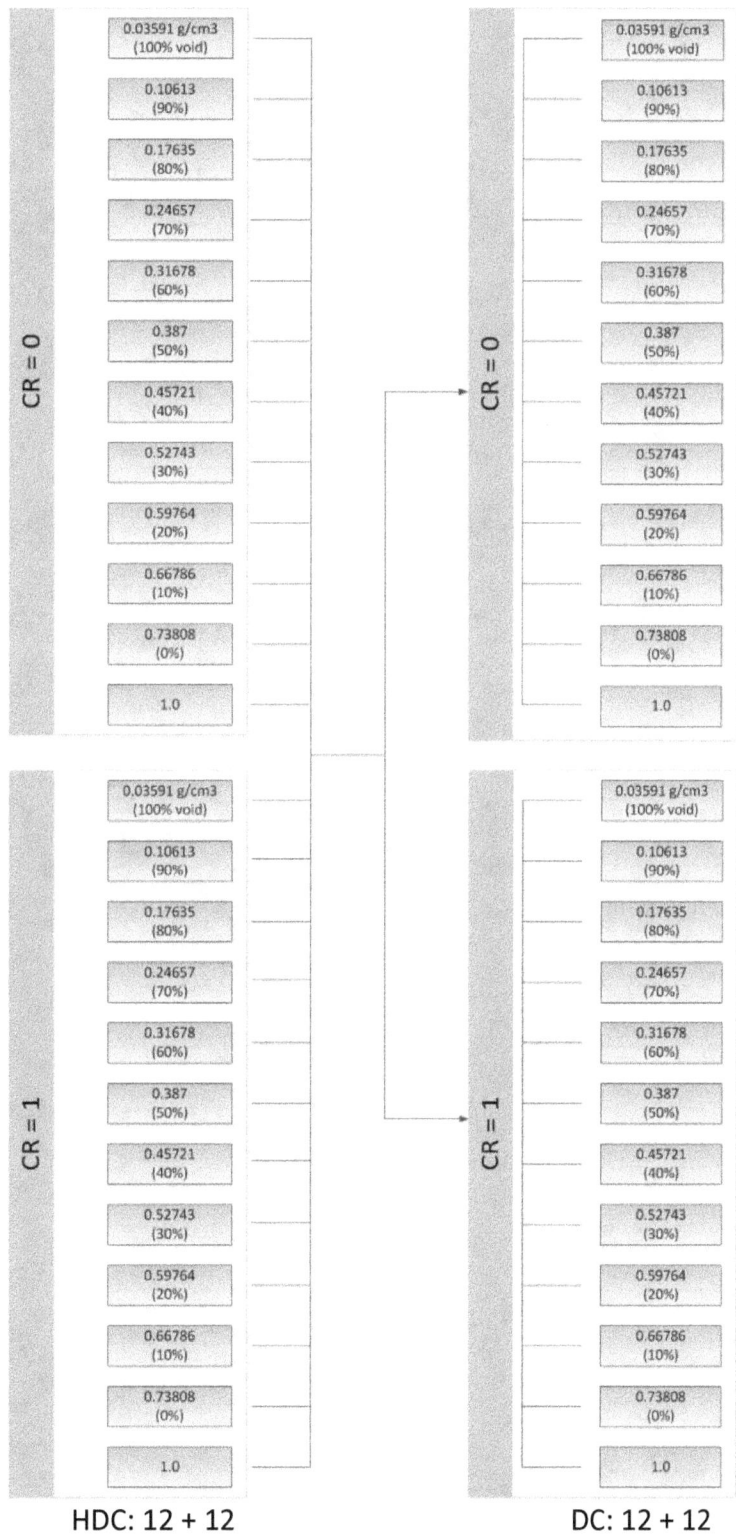

HDC: 12 + 12   DC: 12 + 12

Figure 3. Ref PMAXS histories (HDC) and instantaneous coolant density branches (DC).

3.0   COMPARISONS OF CROSS SECTION SETS

Each of the three XS sets consists of two-group principal macroscopic XS; microscopic XS for Xe/Sm; assembly discontinuity factors; and kinetics parameters, including inverse neutron velocities (1/v), delayed neutron fractions (β), and decay constants of delayed neutrons (λ).

3.1.   COMPARISONS OF CROSS SECTIONS

Two-group macroscopic XS generated with CASMO, HELIOS, and TRITON, as a function of void fraction are shown in Figure 4, Figure 5, and Figure 6, along with the calculated Kinf values.

Figure 4 shows XS comparisons for BOL. BOL fresh fuel has not undergone depletion. Therefore there are no fission products and other isotopes in the fuel.

Figure 5 shows XS as a function of "history void fraction" at EOL (49.57 MWd/kg). The plotted XS are XS at the end of each void history depletion calculation. The horizontal axis is the history void fraction. There are a total of 11 void history cases (0%, 10%, 20%, 30%, 40%, 50%, 60%, 70%, 80%, 90%, and 100%).

Figure 6 shows XS as a function of "branch void fraction" at EOL (49.57 MWd/kg) for the 40% void history. The plotted XS are branch instantaneous XS at the end of the 40% void history depletion. The horizontal axis is the branch void fraction.

Observations on comparisons of XS are summarized below.

1. At BOL, Figure 4 shows that there are about 300 pcm differences in Kinf between TRITON and CASMO. TRITON slightly over predicts Σ_a^2 and $\nu\Sigma_f^2$ at high voids, compared with CAMSO and HELIOS.

2. For history XS comparison, Figure 5 shows that TRITON over predicts Kinf by 200–1500 pcm compared with CASMO or HELIOS. Note that the difference increases as the void fraction increases.

3. The literature reports that the approximations used to speed up CASMO and HELIOS calculations may not be applicable for void fractions higher than 80% (see for example $\nu\Sigma_f^1$ on Figure 5).

4. For branch XS comparisons, overall TRITON over predicts Kinf by 500 pcm compared with CASMO or HELIOS at EOL.

5. TRITON slightly under predicts Σ_{tr}^1 by 2–8%. An investigation indicates that the difference might be due to the collapsing method of transport XS. TRITON currently employs inverse-energy collapse to calculate the two-group transport XS using inverse-energy collapsing with B1-critical spectrum correction; however, both CASMO and HELIOS use direct energy collapsing.

6. The values Σ_a^1 and $\nu\Sigma_f^1$ are the two key parameters for the study of XS sensitivity. Both XS curves show more curvature at high void than low void.

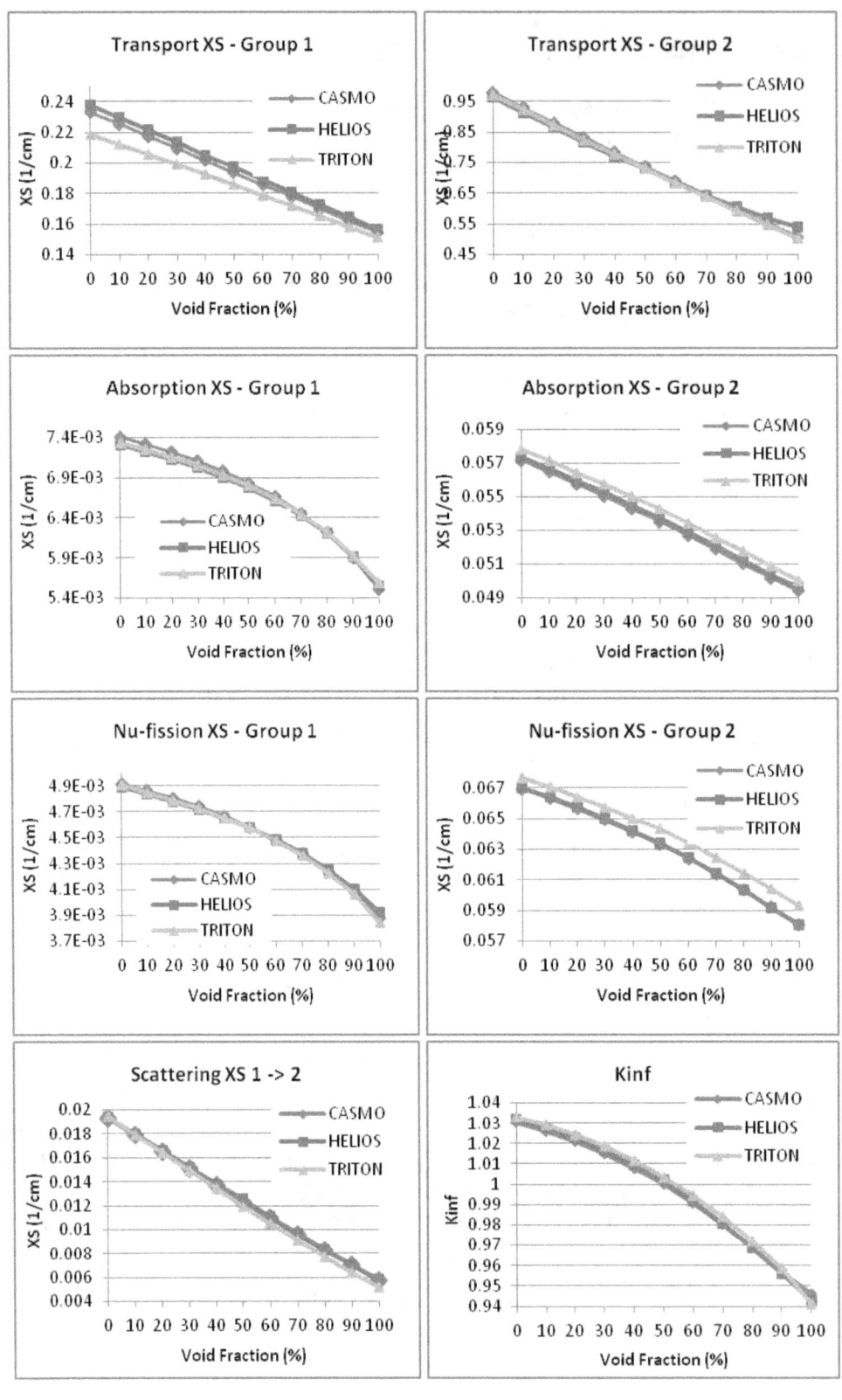

Figure 4. Cross Section Comparisons at BOL.

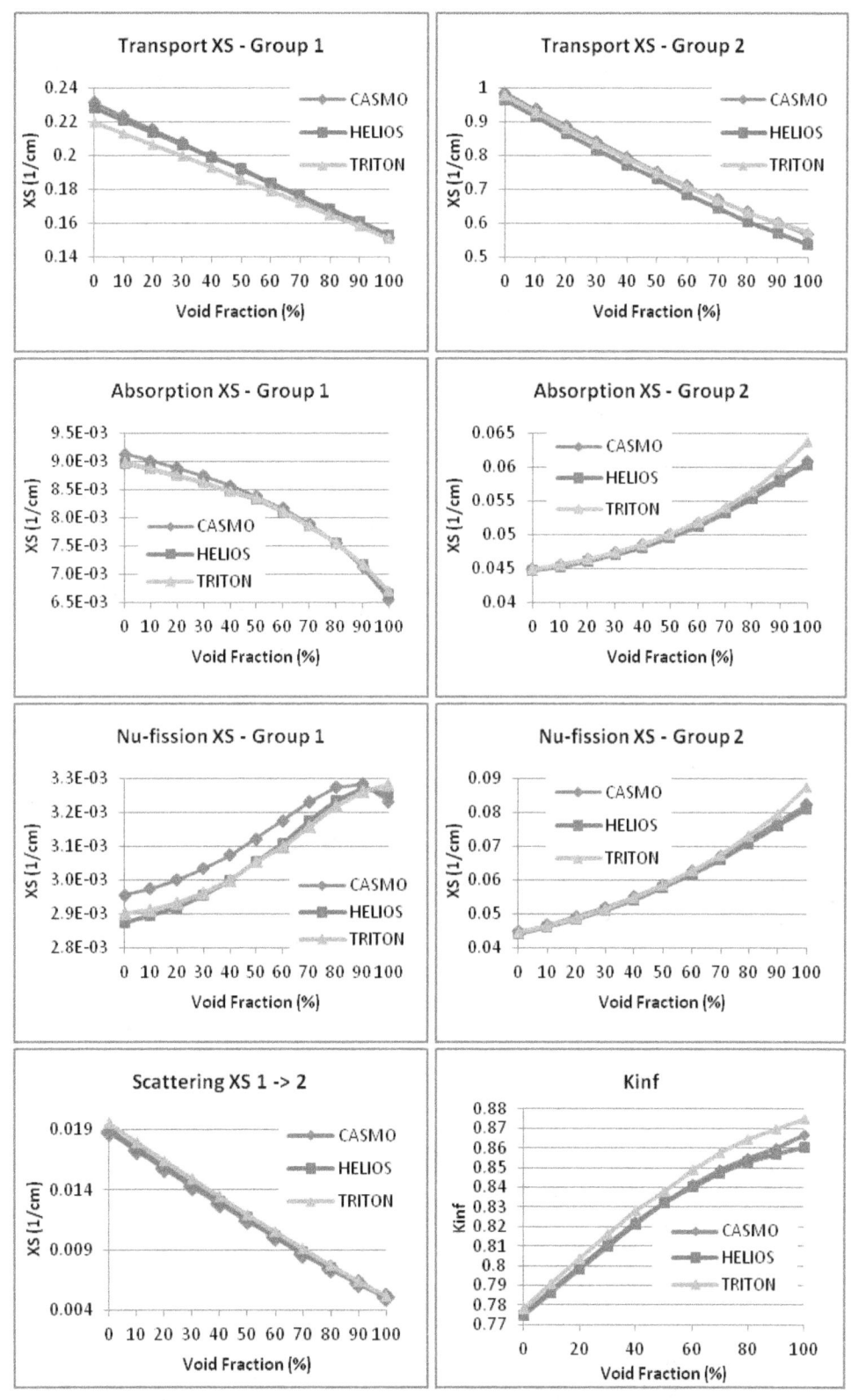

Figure 5. HDC Cross Section Comparisons at EOL.

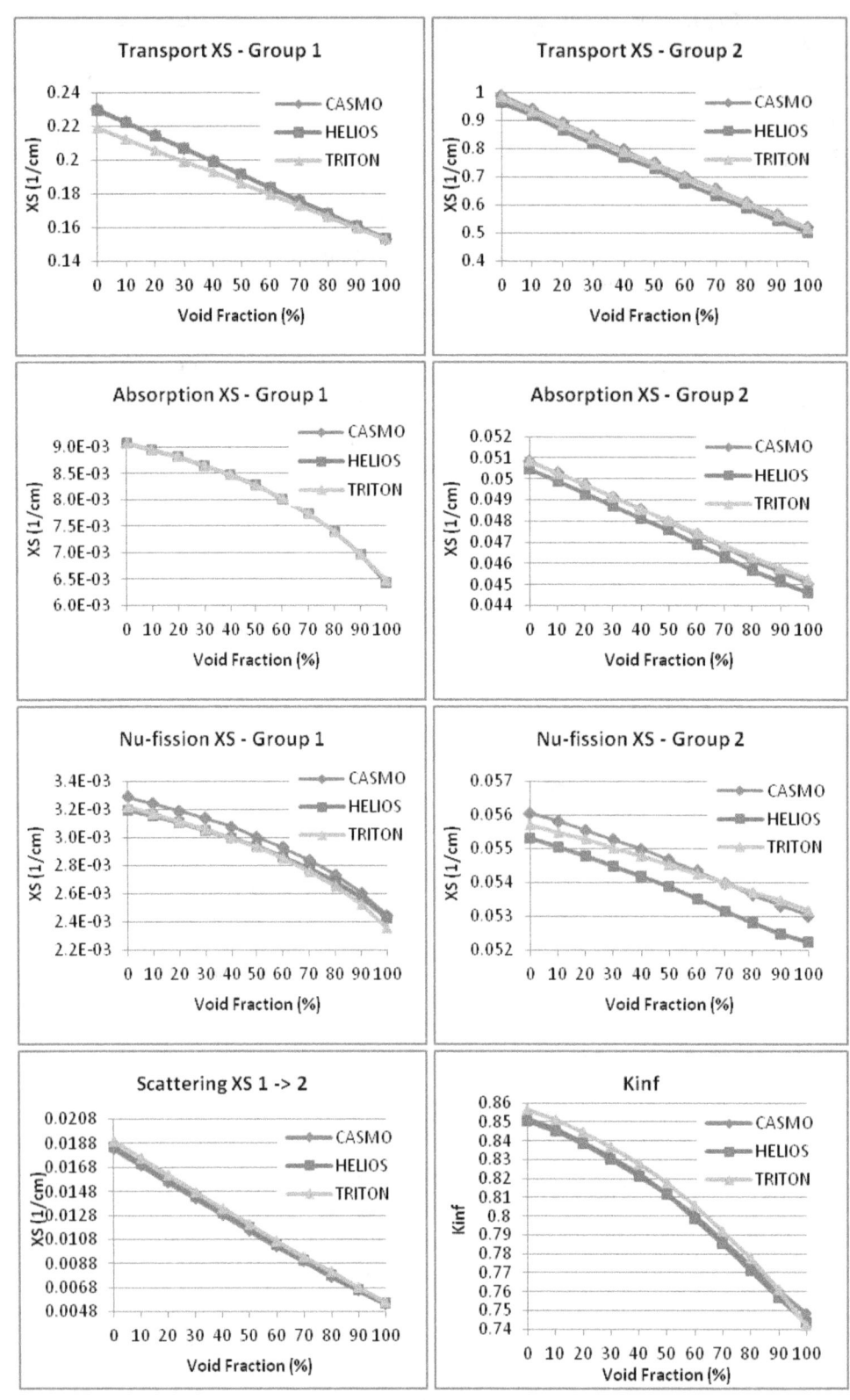

Figure 6. Branch Cross Section at EOL (40% void history CR = 0).

3.2. BENCHMARK BETWEEN TRITON AND KENO

In general, as the void fraction increases, the neutron spectrum becomes harder; so XS databases plus solution methods may not be applicable. Therefore, there is some concern about the predictability of lattice codes for fuel regions with very high void. Our study found that CASMO and HELIOS have convergence difficulties for the void fraction >80%. To investigate how well TRITON performs at high void, a benchmark was performed between TRITON and KENO for a 7×7 BWR bundle similar to the PB type-3c, as shown in Figure 7. On the left plot of the figure, the green line is the results of the TRITON 49 group XS library, the red line is the results of the TRITON 238 group XS library, and the KENO results are shown with the blue line. The right plot shows relative differences between TRITON and KENO. The 49G TRITON results diverge at 100% void; the 238G results do not.

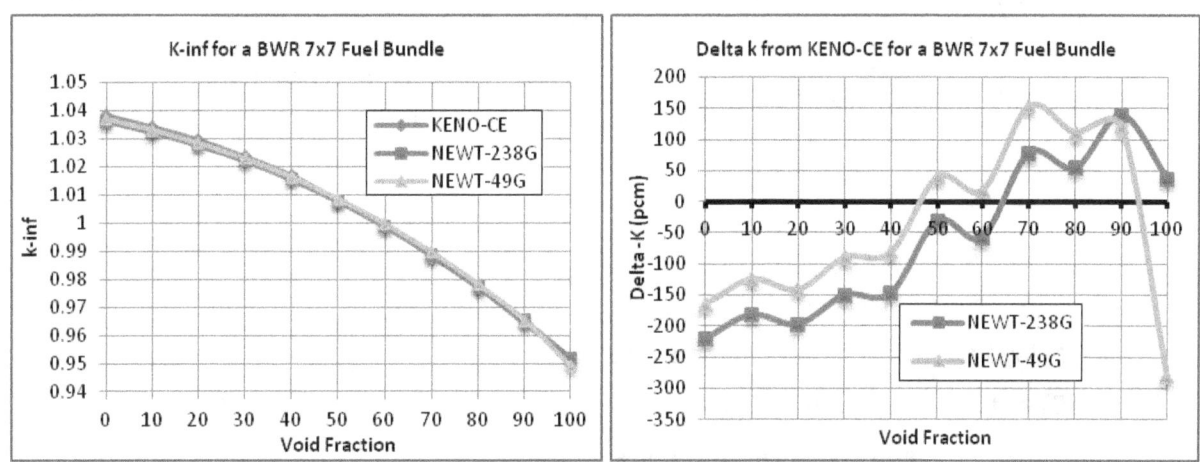

Figure 7. Benchmark between TRITON and KENO.

3.3. COMPARISONS OF KINETICS PARAMETERS

Kinetics data, together with two-group XS, are generated with lattice calculations. The kinetics data used by PARCS include two-group inverse neutron velocities (1/v), delayed neutron fractions (β), and decay constants of delayed neutrons (λ).

3.3.1. 1/v

In CASMO, the assembly average value of the inverse neutron velocity is calculated by three different collapsing schemes:

$$\left(\frac{1}{v}\right)_G = \frac{\sum_{g \in G} \phi_g (1/v)_g}{\sum_{g \in G} \phi_g} \tag{3.1}$$

$$\left(\frac{1}{v}\right)_G^* = \frac{\sum_{g \in G} \phi_g^* (1/v)_g}{\sum_{g \in G} \phi_g^*} \tag{3.2}$$

$$\left(\frac{1}{v}\right)_G' = \frac{\sum_{g \in G} \phi_g^* (1/v)_g \phi_g}{\sum_{g \in G} \phi_g^*} \tag{3.3}$$

11

In TRITON and HELIOS, Eq. (3.1) is employed to collapse two-group inverse neutron velocities. A comparison of inverse neutron velocities between TRITON and CASMO shows that the agreement is poor for high void histories; for example, the relative difference may be up to 30% for the fast group. HELIOS values have very good agreement with CAMSO for the fast group but are more than 20% lower than the CAMSO values for the thermal group.

3.3.2. β_{eff}

Traditionally, β_{eff} is calculated using bilinear weighting as below.

$$\beta_{eff,i} = \frac{\sum_m \int dr \int dE \chi_{d,i}^m(E) \phi^*(r,E) \int dE' \beta_i^m v \Sigma_f^m \phi(r,E')}{\sum_m \int dr \int dE \chi_t^m(E) \phi^*(r,E) \int dE' v \Sigma_f^m \phi(r,E')} = \frac{\sum_m \beta_i^m \int dr \int dE \chi_{d,i}^m(E) \phi^*(r,E) \int dE' v \Sigma_f^m \phi(r,E')}{\sum_m \int dr \int dE \chi_t^m(E) \phi^*(r,E) \int dE' v \Sigma_f^m \phi(r,E')}, i = 1,..,6 \qquad (3.4)$$

where

    m = the mth isotope

    $\phi^*(r,E)$ = adjoint flux

    $\phi(r,E)$ = neutron flux

    β_i = delayed neutron fraction of the ith group

    $\chi_{d,i}^m(E)$ = delayed neutron spectrum of the ith group

    $\chi_t^m(E)$ = total fission spectrum (prompt + delayed)

    $v \Sigma_f \phi(r,E')$ = total fission neutrons (prompt + delayed)

If $\phi^*(r, E) = constant \ (or \ 1)$, we obtain fission weighted β as

$$\beta_i = \frac{\sum_m \beta_i^m \int dr \int dE' v \Sigma_f^m \phi(r,E')}{\sum_m \int dr \int dE' v \Sigma_f^m \phi(r,E')}, i = 1, ...,6 \qquad (3.5)$$

In HELIOS, β_{eff} is calculated with K-ratio weighting [2], which is calculated as

$$\beta_{eff} = \frac{\langle \chi_d v_d \rangle}{\langle \chi v \rangle} \approx 1 - \frac{\langle \chi_p v_p \rangle}{\langle \chi v \rangle} = 1 - \frac{k_p}{k} \quad , \qquad (3.6)$$

where k_p is calculated based on the prompt neutron spectrum. It is found that β_{eff} calculated using K-ratio weighting can produce nearly identical results to those calculated using the adjoint weighting method [2].

Delayed neutron yields for major isotopes used in TRITON are given in Table 3. The values used by CASMO and HELIOS are slightly different, varying between 2% to 10% differences for different isotopes.

Table 3. TRITON delayed neutron yields [3]

	^{235}U		^{238}U		^{239}Pu		^{240}Pu		^{241}Pu	
	Group 1	Group 2	Group 1	Group 2	Group 1	Group 2	Group 1	Group 2	Group 1	Group 2
β_m	0.0064	0.0067	0.0164	0	0.002	0.0022	0.0029	0	0	0.0054
$\alpha_{m\,i}$										
1	0.038	0.033	0.013		0.038	0.035	0.028			0.01
2	0.213	0.219	0.137		0.28	0.298	0.273			0.229
3	0.188	0.196	0.162		0.216	0.211	0.192			0.173
4	0.407	0.395	0.388		0.328	0.326	0.35			0.39
5	0.128	0.115	0.225		0.103	0.086	0.128			0.182
6	0.026	0.042	0.075		0.035	0.044	0.029			0.016

Figure 8 shows comparisons of fission-neutron-weighted β (calculated with Eq. [3.3]). Poor agreement between TRITON and CASMO (or HELIOS) occurs because these calculations were performed with V6.1 of SCALE, and TRITON did not include the ν term in Eq. (3.3) when calculating fission-weighted β. In other words, V6.1 TRITON calculated the "fission-weighted" beta, while CASMO calculates the "fission-neutron-weighted" beta. As a result of this work, TRITON now includes an option to include the ν term, which is the recommended setting. A new option has been added in SCALE V6.1.1, and TRITON can also compute the fission-neutron-weighted beta, which accounts for the differences observed between the codes. Figure 9 shows an example TRITON calculation.

Figure 10 shows CASMO fission-neutron-weighted β and adjoint infinite flux–weighted β. It is seen that adjoint flux–weighted β is about 10% less than fission-neutron-weighted β.

In general, β decreases with burnup because of the depletion of ^{235}U and the production of ^{329}Pu, which has a smaller β. At high voids, β increases because of enhanced ^{238}U fission, which has higher β than ^{235}U and ^{329}Pu.

Figure 11 shows HELIOS fission-weighted β and K-ratio–weighted β. Generally, k-ratio–weighted β can be up to 20% less than fission-neutron-weighted β for unrodded assemblies (CR =0). The right plot in Figure 11 shows that K-ratio–weighted β can be 20% less for CR =1 than for CR =0.

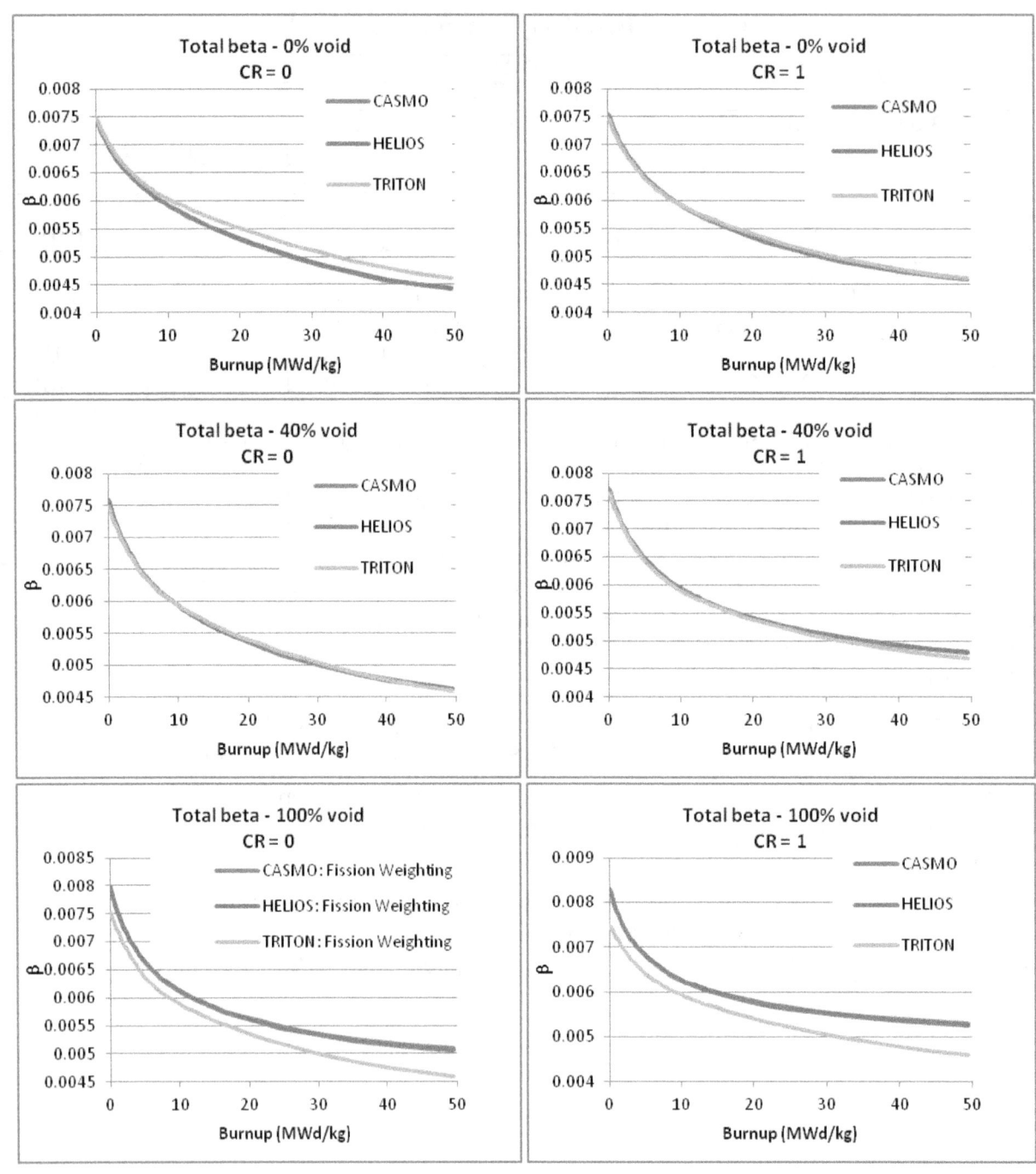

Figure 8. Fission-neutron-weighted β.

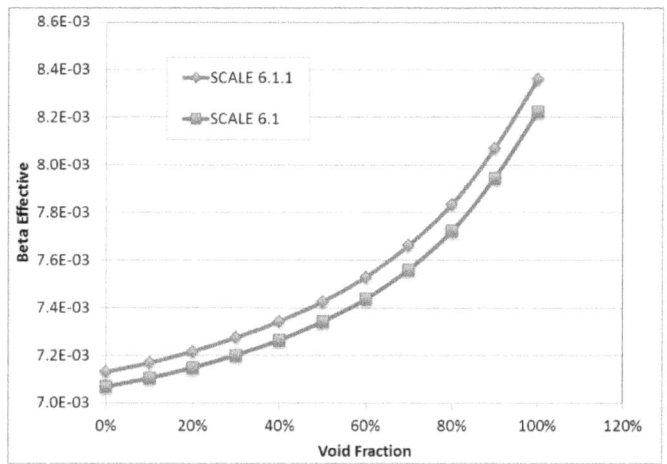

Figure 9. Fission-weighted (6.1) vs neutron-fission-weighted (6.1.1) beta effective.

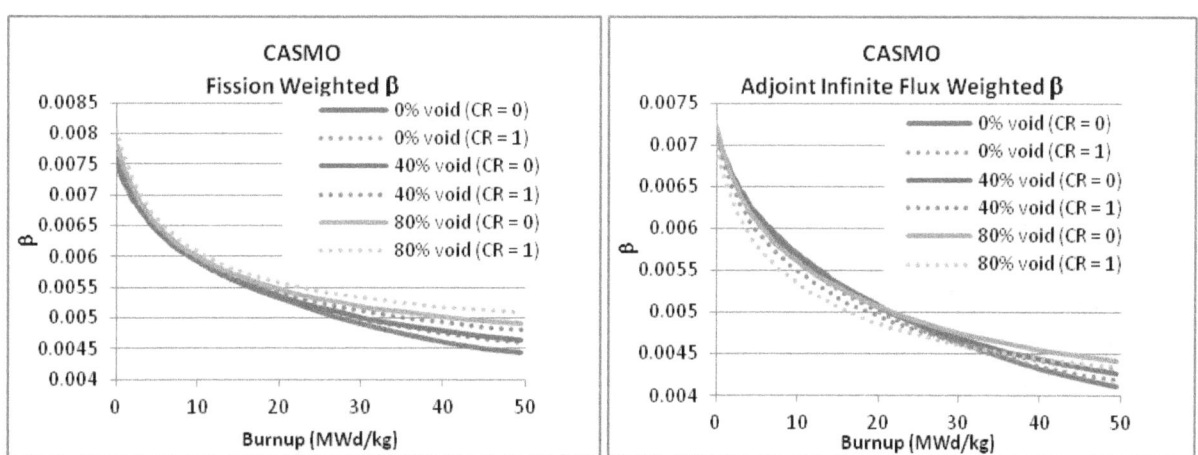

Figure 10. CASMO fission-neutron-weighted β and adjoint infinite flux–weighted β.

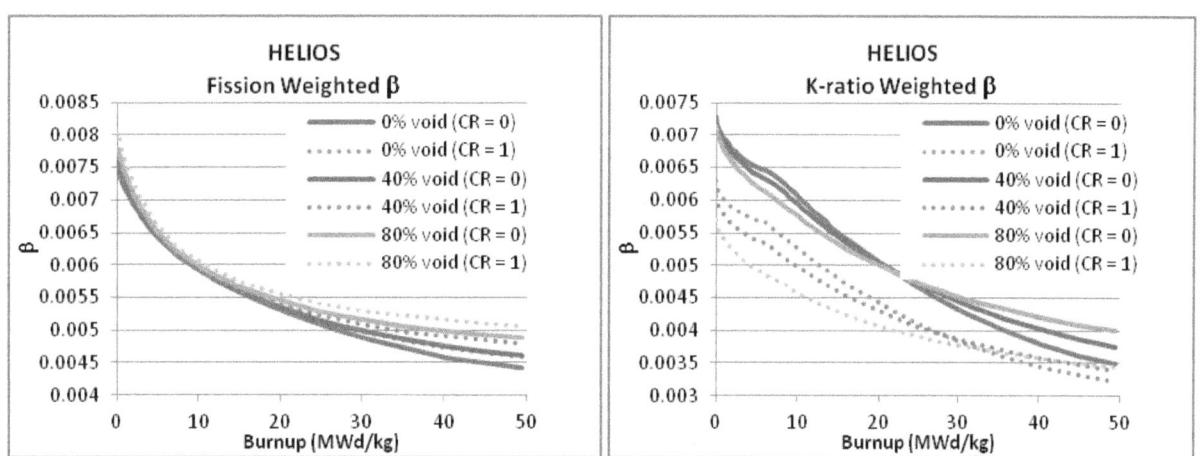

Figure 11. HELIOS fission-neutron-weighted β and K-ratio–weighted β.

4.0 SENSITIVITY ANALYSIS OF CROSS SECTIONS

Three TRACE/PARCS models employed to conduct the XS sensitivity study are

1. Single CHAN model. The TRACE model consists of a CHAN, an inlet BREAK, and an outlet BREAK. The TRACE model is coupled with the PARCS core model. The flow is determined by the differential pressure between the inlet and outlet BREAKs. The transient is initiated by a pressure perturbation at the outlet. The power and flow of the single-CHAN model represent thermal-hydraulic conditions in a typical BWR.

2. Oskarshamn plant model. The TRACE model is a detailed representation of the Oskarshamn plant, including the reactor pressure vessel (RPV), steam line, recirculation loop, and so on. The core consists of 444 fuel assemblies, modeled with 222 CHAN components in half-core symmetry. Each CHAN has 25 non-uniform powered axial nodes. The reactor has a power of 1723.8 MWth and a core flow of 3210 kg/s. The PARCS model has 444 radial nodes (node/assembly) and 25 axial powered nodes with uniform 12.42 cm spacing. A core-wide stability transient is induced by control rod perturbation.

3. Ringhals plant model. The TRACE model is a detailed representation of the Ringhals plant, including the RPV, steam line, recirculation loop, and so on. The core consists of 648 fuel assemblies, modeled with 325 CHAN components in half-core symmetry. Each CHAN has 25 powered axial nodes with uniform 14.72 cm spacing. The reactor has a power of 1770.6 MWth and a core flow of 3665 kg/s. The PARCS model has 648 radial nodes (node/assembly) and 25 axial powered nodes with uniform 14.72 cm spacing. A core-wide stability transient is induced by control rod perturbation.

The PMAXS data for PB type 3c fuel were used for all fuel types in these PARCS core models. Sensitivity analysis results for HDC effects and branch effects are presented in the following sections. Note that the results presented were calculated with the CASMO PMAXS data.

4.1. DC HISTORY EFFECTS

The CASMO PMAXS data consist of 12 uncontrolled HDCs (CR = 0) and 12 controlled HDCs (CR = 1), each history having 12 DC branches as shown in Figure 3. The DC ranges from 0.03591 g/ cm^3 (100% void) up to 0.73808 g/ cm^3 (0% void), with 10% spacing. One additional density at 1.0 g/ cm^3 is for cold, unpressurized, core conditions.

Macroscopic XS have a strong dependence on the history effects of DC and control rod position. Figure 12 shows Σ_a^1 and $\nu\Sigma_f^1$ as a function of history void fraction at BOL and EOL, respectively. The black dashed line on each figure is the second-order polynomial fitting of the XS at 0, 40, and 70% void.

Note that both Σ_a^1 and $\nu\Sigma_f^1$ curves show more nonlinearity at high void fraction (>70%) than at low void fraction. Thus the extrapolation of the XS from the 70% void may overestimate the XS at high void faction.

For the study of HDC effects, the PMAXS file has been broken down into five PMAXS files as shown in Figure 13 and Figure 14. These small PMAXS files are described as follows.

1. HDC 12+12: This is the reference case with 12 rodded and 12 unrodded HDCs. Each history has 12 rodded and 12 unrodded DC branches.

2. HDC 7+7: Seven rodded and seven unrodded HDCs (0%, 20%, 40%, 60%, 80%, 100% void, and 1 g/cc, for both CR = 0 and 1). Each history has 12 rodded and 12 unrodded DC branches.

3. HDC 5+5: Five rodded and five unrodded HDCs (0%, 40%, 80%, 100% void, and 1 g/cc). Each history has 12 rodded and 12 unrodded DC branches.

4. HDC 4+4: Four rodded and four unrodded HDCs (0%, 40%, 80% void, and 1 g/cc, for both CR = 0 and 1). Each history has 12 rodded and 12 unrodded DC branches.

5. HDC 3+3: Three rodded and 3 unrodded HDCs (0%, 40%, 80% void, for both CR = 0 and 1). Each history has 12 rodded and 12 unrodded DC branches.

6. HDC 3+1: One rodded and three unrodded HDCs (0%, 40%, 80% void for CR = 0, and 40% for CR =1). Each history has 12 rodded and 12 unrodded DC branches.

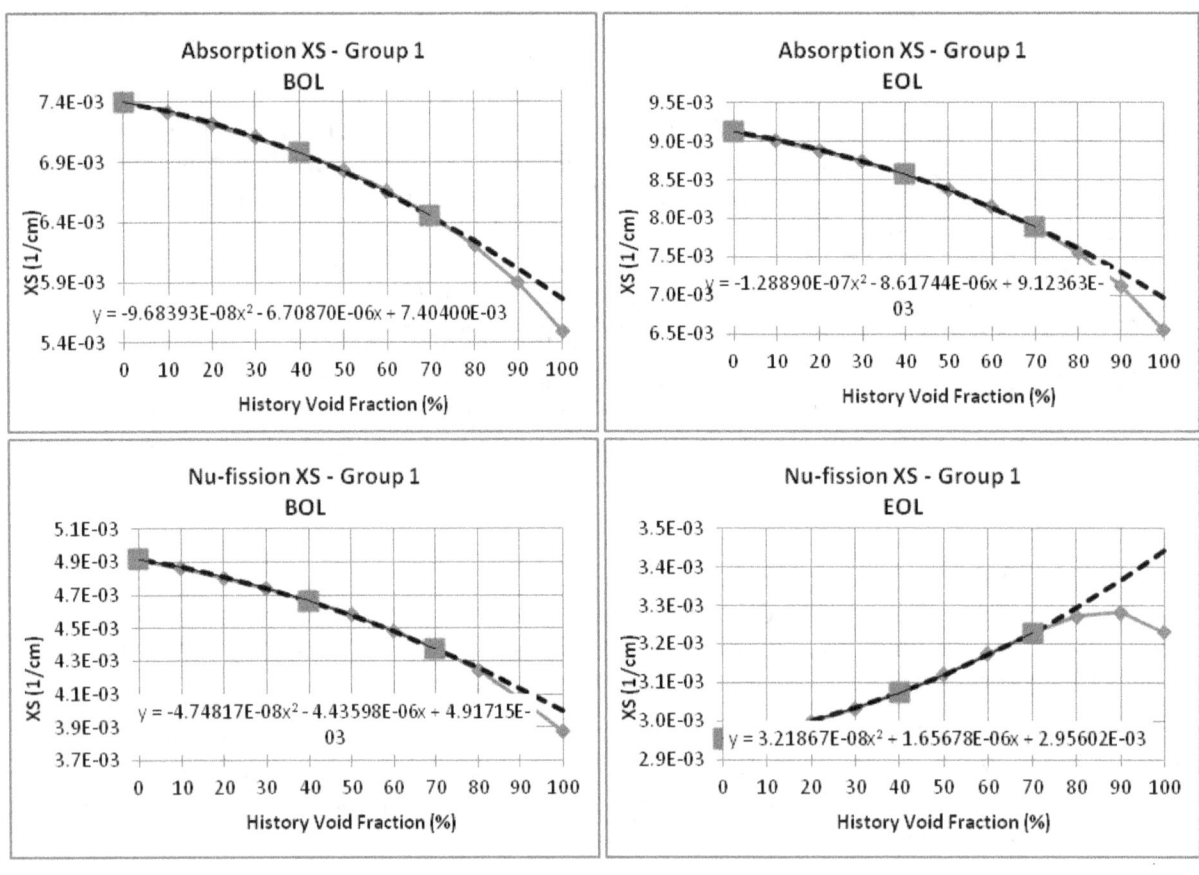

Figure 12. Σ_a^1 and $\nu\Sigma_f^1$ at BOL and EOL.

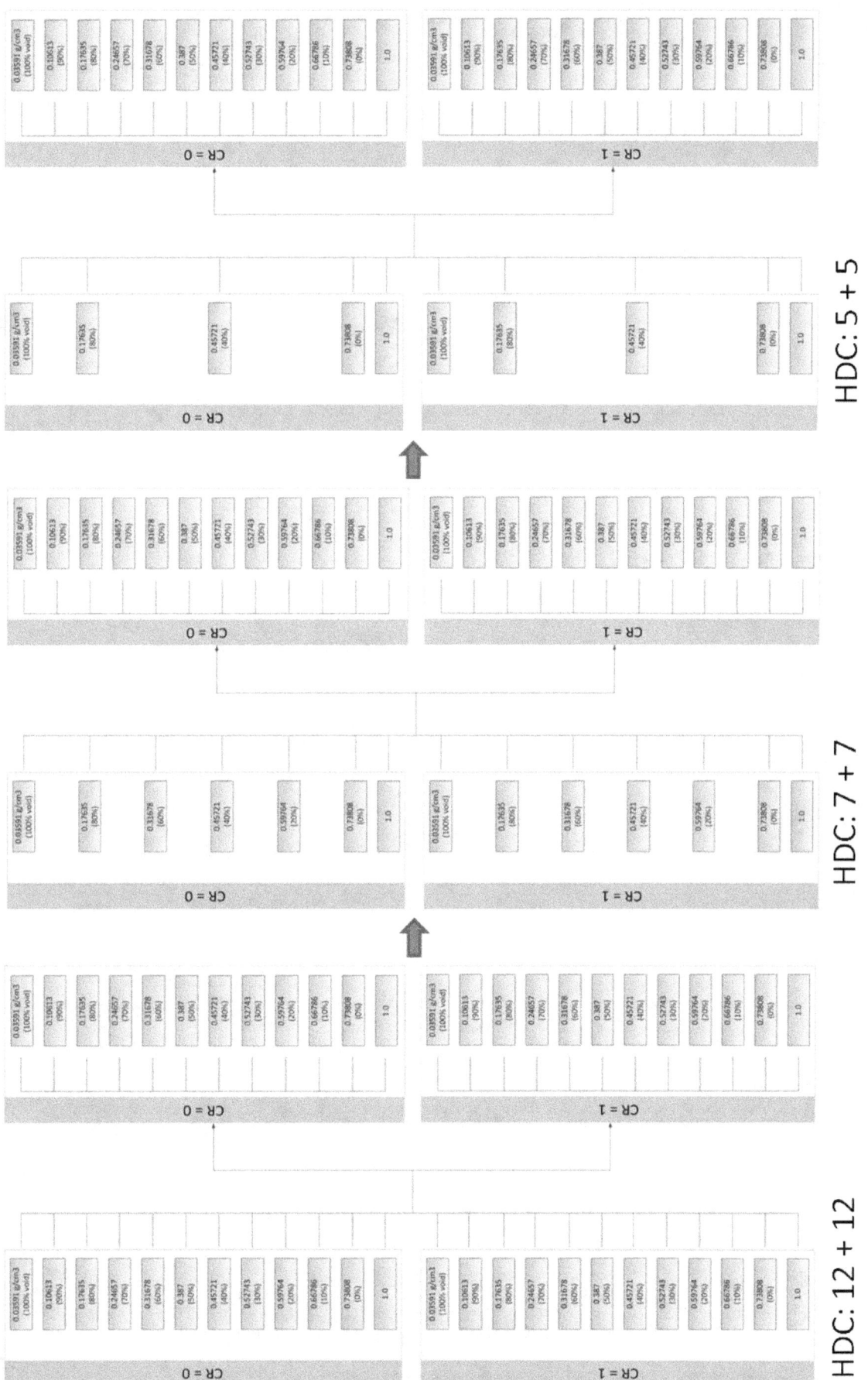

HDC: 5 + 5

HDC: 7 + 7

HDC: 12 + 12

Figure 13. PMAXS HDC breakdown.

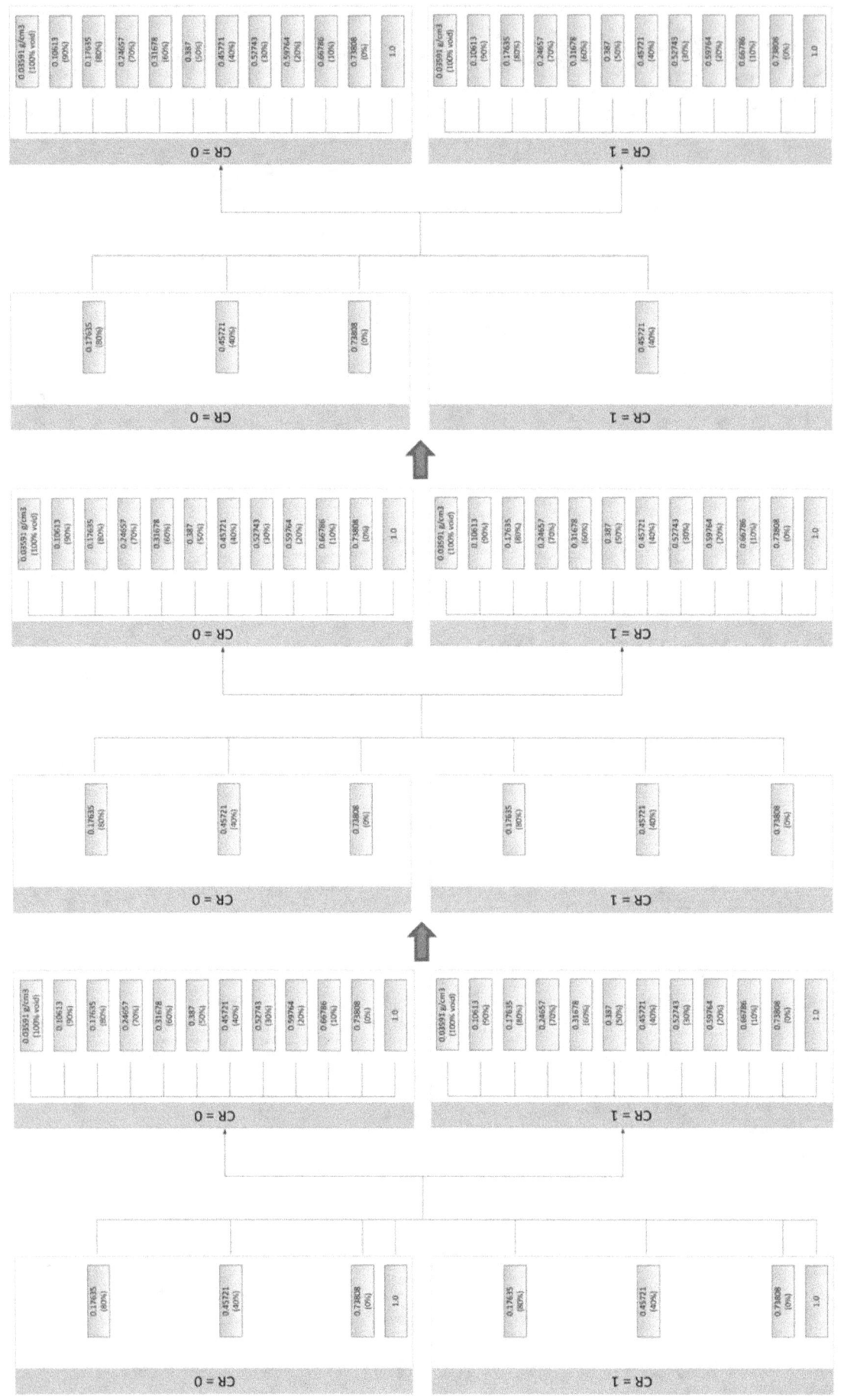

HDC: 3 + 1

HDC: 3 + 3

HDC: 4 + 4

Figure 14. PMAXS HDC breakdown (cont.).

4.1.1. Results of Single CHAN Model

Steady-state and transient results calculated for the three TRACE/PARCS models are presented in this section. All simulations were run with TRACE V5.509.

HDC and HCR inputs for this case are shown in Figure 15. On the top of the channel, the coolant has an HDC of about 0.1 g/cc (i.e., 90% void). Steady-state results for each HDC case are summarized in Table 4, and they include core eigenvalue (K_{eff}), 3-dimensional power peaking factor (P_{xyz}), core radial assembly power peaking factor (P_{xy}), and core axial power peaking factor (P_z). The value of $P_{xy} = 1$ because there is only one fuel assembly in the core. Figure 16 shows power transient oscillations induced by a pressure perturbation at the outlet pressure boundary. Overall, all history cases predict almost the same results for the steady-state and transient cases.

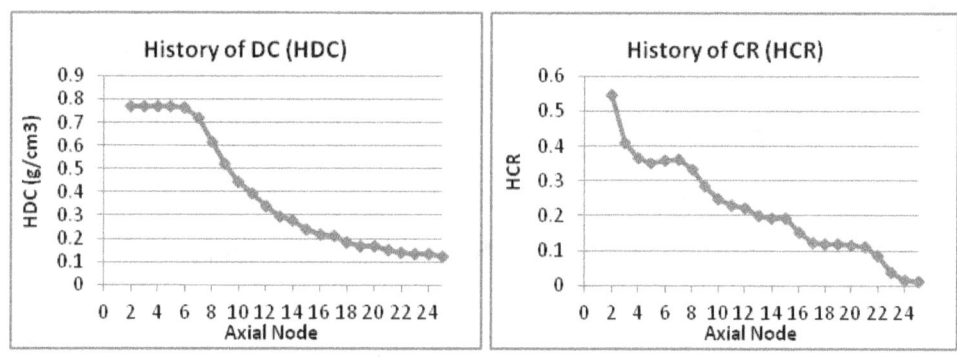

Figure 15. HDC and HCR input for single-CHAN model.

Table 4. Steady-state parameters

Steady-state						
Parameters	**HDC: 12+12**	**HDC: 7+7**	**HDC: 5+5**	**HDC: 4+4**	**HDC: 3+3**	**HDC: 3+1**
Keff	0.999154	0.999170	0.999218	0.999208	0.999208	0.999173
Pxyz	1.773	1.773	1.773	1.773	1.773	1.770
Pxy	1.0	1.0	1.0	1.0	1.0	1.0
Pz	1.773	1.773	1.773	1.773	1.773	1.770

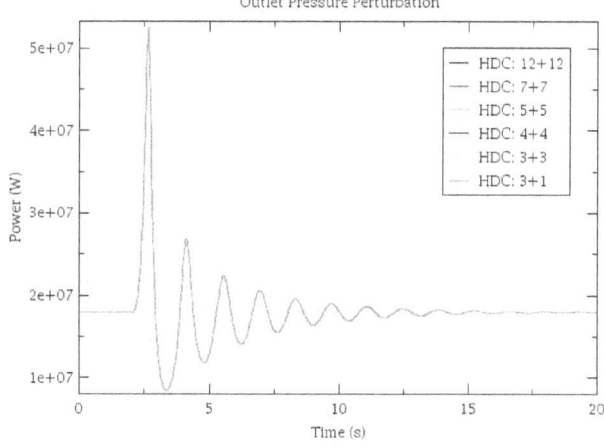

Figure 16. Response to a pressure perturbation for single-CHAN model.

21

4.1.2. Results of Oskarshamn Core-wide Stability

Figure 17 shows HDC and HCR inputs for this case for the different channels modeled. On the top of the channel, the coolant has an average HDC of about 0.2 g/cc (i.e., 75% void). Steady-state results for each history case are summarized in Table 5. There are no appreciable differences between HDC cases. However, the case of HDC 7+7 has the best agreement with the reference case HDC 12 + 12. Figure 18 shows core transient flow oscillations induced by control rod perturbation. Overall, all history cases predict almost the same results for the steady-state and transient.

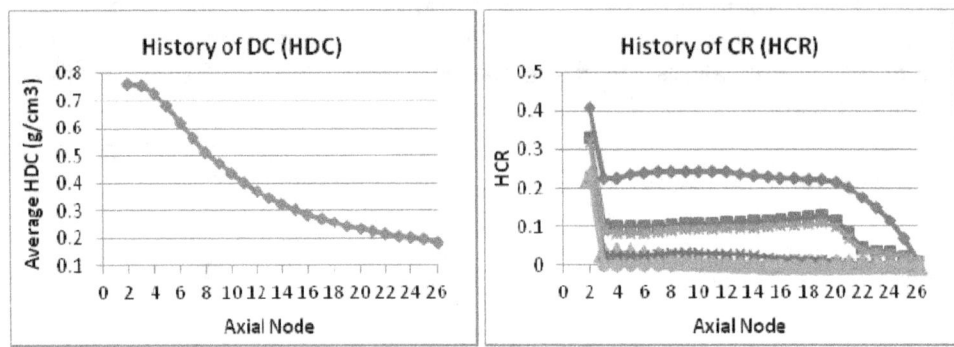

Figure 17. HDC and HCR input for Oskarshamn

Table 5. Steady-state parameters for Oskarshamn

Steady-state						
Parameters	HDC: 12+12	HDC: 7+7	HDC: 5+5	HDC: 4+4	HDC: 3+3	HDC: 3+1
Keff	0.901037	0.901057	0.90119	0.901182	0.901181	0.900942
Pxyz	2.139	2.138	2.128	2.128	2.129	2.156
Pxy	1.482	1.482	1.482	1.481	1.481	1.481
Pz	1.315	1.314	1.307	1.307	1.308	1.333

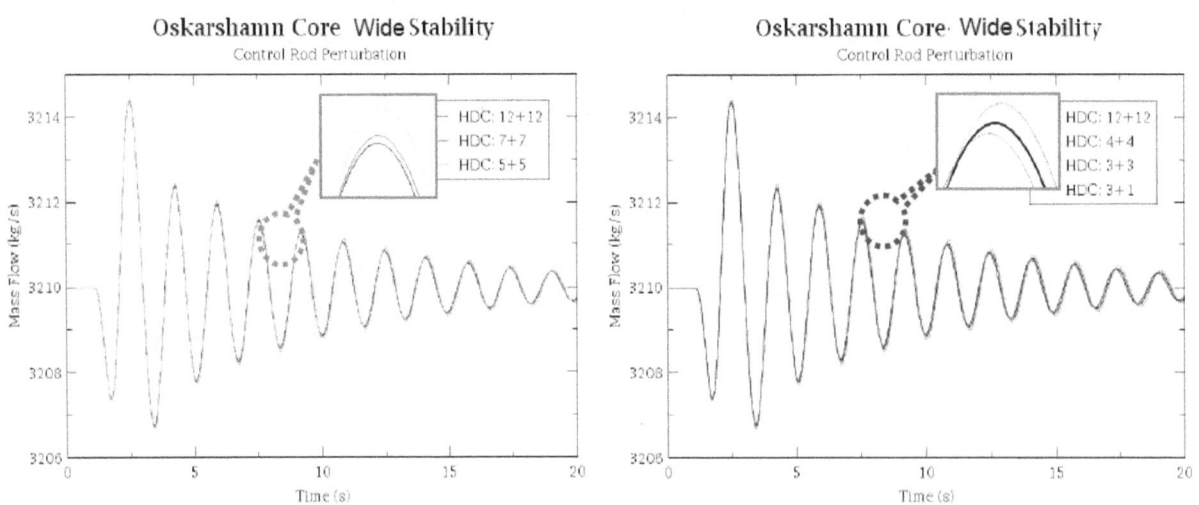

Figure 18. Response to a control rod perturbation.

4.1.3. Results of Ringhals Core-wide Stability

HDC and HCR inputs for this case are shown in Figure 19. On the top of the channel, the coolant has an average HDC of about 0.2 g/cc (i.e., 75% void). Steady-state results for each history case are summarized in Table 6. Again, there are no appreciable differences between the history cases. The case of HDC 7+7 has best agreement with the reference case HDC 12+12. Figure 20 shows transient core flow stability induced by control rod perturbation. Again, all history cases predict almost the same results for the steady-state and transient.

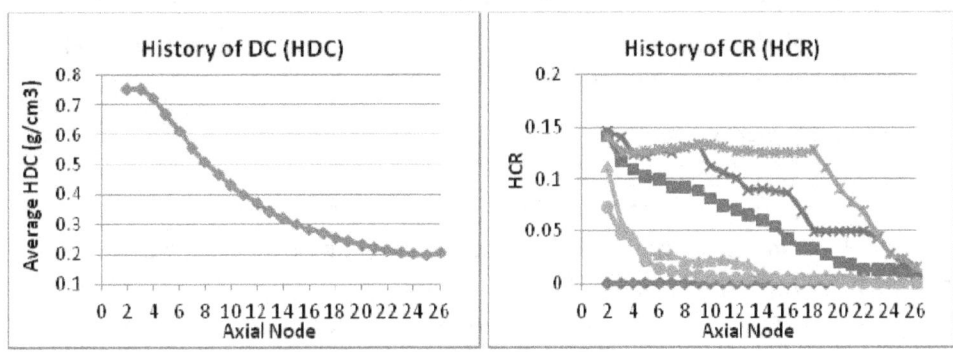

Figure 19. HDC and HCR input for Ringhals.

Table 6. Steady-state parameters for Ringhals

	Steady-state					
Parameters	HDC: 12+12	HDC: 7+7	HDC: 5+5	HDC: 4+4	HDC: 3+3	HDC: 3+1
Keff	0.925303	0.925328	0.925479	0.925469	0.925468	0.925368
Pxyz	2.388	2.386	2.373	2.374	2.375	2.378
Pxy	1.688	1.688	1.688	1.688	1.688	1.686
Pz	1.164	1.163	1.166	1.166	1.166	1.163

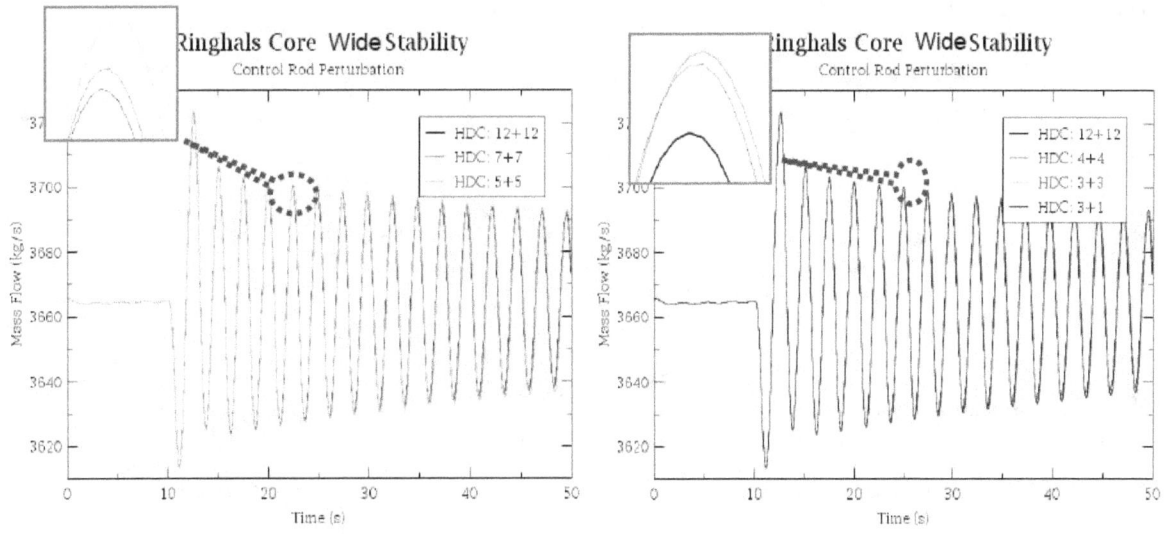

Figure 20. Core flow stability for Ringhals.

4.1.4. Conclusion of DC History Effects

Based on the above analysis of the HDC effects, the following conclusions can be reached:

1. BWR steady-state variables (i.e., K_{eff}, P_{xyz}, P_{xy}, and P_z) are not very sensitive to the HDC and HCR inputs (history effects). Results of HDC 7+7 (20% void spacing) have the best agreement with the reference case HDC 12+12.

2. BWR transients are not highly sensitive to the HDC and HCR effects. Results of HDC 7+7 (20% void apart) have best agreement with the reference case HDC 12+12.

3. Even the case of HDC 3+1 (3 uncontrolled HDC and 1 controlled HDC) does not show any significant deviation from the reference case.

4.2. DC BRANCH EFFECTS

Figure 21 shows Σ_a^1 and $\nu\Sigma_f^1$ as a function of branch void faction at BOL and EOL. The black dashed line on each figure is the second-order polynomial fitting of the cross sections at 0, 40, and 70% void. At high void fraction (>70%) both Σ_a^1 and $\nu\Sigma_f^1$ curves show more nonlinearity (i.e., curvature) than those at low void fractions. Therefore the extrapolation of the branch XS from the 70% void may overestimate the XS at high void faction.

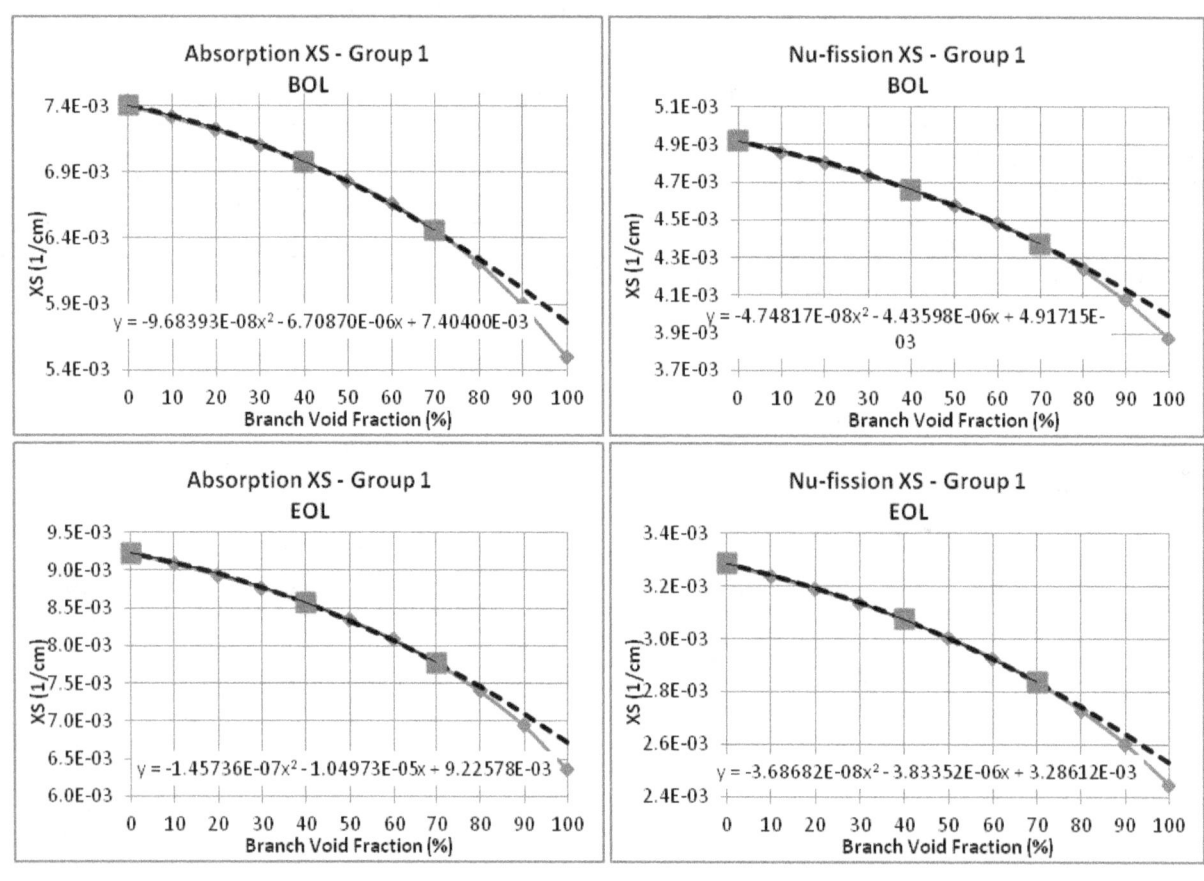

Figure 21. Σ_a^1 and $\nu\Sigma_f^1$ at BOL and EOL as function of branch void.

For the study of DC branch effects, the PMAXS file has been broken down into five PMAXS files with less branches. These reduced PMAXS files are described as follows.

1. DC 12+12: This is the reference case with seven rodded and seven unrodded HDCs (0%, 20%, 40%, 60%, 80%, 100% void, and 1 g/cm^3, for both CR = 0 and 1). Each history has 12 rodded and 12 unrodded DC branches (0%, 10%, 20%, 30%, 40%, 50%, 60%, 70%, 80%, 90%, 100% void, and 1 g/cc, for both CR = 0 and 1).

2. DC 7+7: Seven rodded and seven unrodded HDCs. Each history has seven rodded and seven unrodded DC branches (0%, 20%, 40%, 60%, 80%, 100% void, and 1 g/cm^3, for both CR = 0 and 1).

3. DC 5+5: Seven rodded and seven unrodded HDCs. Each history has five rodded and five unrodded DC branches (0%, 40%, 80%, 100% void, and 1 g/cm^3, for both CR = 0 and 1).

4. DC 4+4: Seven rodded and seven unrodded HDCs. Each history has four rodded and four unrodded DC branches (0%, 40%, 80% void, and 1 g/cc, for both CR = 0 and 1).

5. DC 3+3: Seven rodded and seven unrodded HDCs. Each history has 12 and 12 unrodded DC branches (0%, 40%, 80% void, for both CR = 0 and 1).

Steady-state and transient results calculated for the three TRACE/PARCS models are presented in the following sections.

4.2.1. Results for Single CHAN Model

Steady-state results for each DC branch case are summarized in Table 7. Figure 22 shows the steady-state core axial DC. The coolant in the top fuel channel is above 80% void. Figure 23 shows the power transient induced by an outlet pressure perturbation. It shows clearly that the cases of DC 4+4 and DC 3+3 differ significantly from the reference case. It indicates that the 100% void branch is necessary to accurately interpolate XS for the core region with a void fraction greater than 80%. In other words, XS extrapolation from 80% cannot accurately generate XS for the core upper region with the void fraction > 80%.

In addition, the results of the case of DC 4+4 are almost identical to those of DC 3+3. For this simulation, the core inlet coolant is subcooled with the density of 0.77 g/cc, which is higher than the saturated density of 0.738 g/cm^3 for 0% void. For the case DC 3+3, the XS for the inlet DC has to be extrapolated from the branch of 0.738 g/cm^3 (0% void). However, the case of DC 4+4 has a XS branch at 1 g/cm^3, so the XS for the inlet DC is interpolated between 1 g/cm^3 and 0.73808 g/cm^3 for this case. Good agreement between the cases DC 4+4 and 3+3 means that XS extrapolation is good at the high-density end (the branch XS curve at high density is more linear than at low density).

Table 7. Steady-state parameters for single-CHAN model

Parameters	Steady-state				
	DC: 12+12	DC: 7+7	DC: 5+5	DC: 4+4	DC: 3+3
Keff	0.99917	0.999342	0.999729	0.999021	0.999021
Pxyz	1.773	1.772	1.783	1.794	1.794
Pxy	1	1	1	1	1
Pz	1.773	1.772	1.783	1.794	1.794

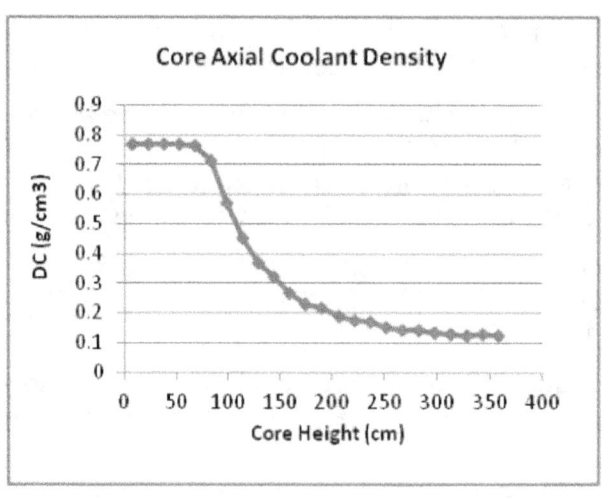

Figure 22. Axial coolant density for single-CHAN model.

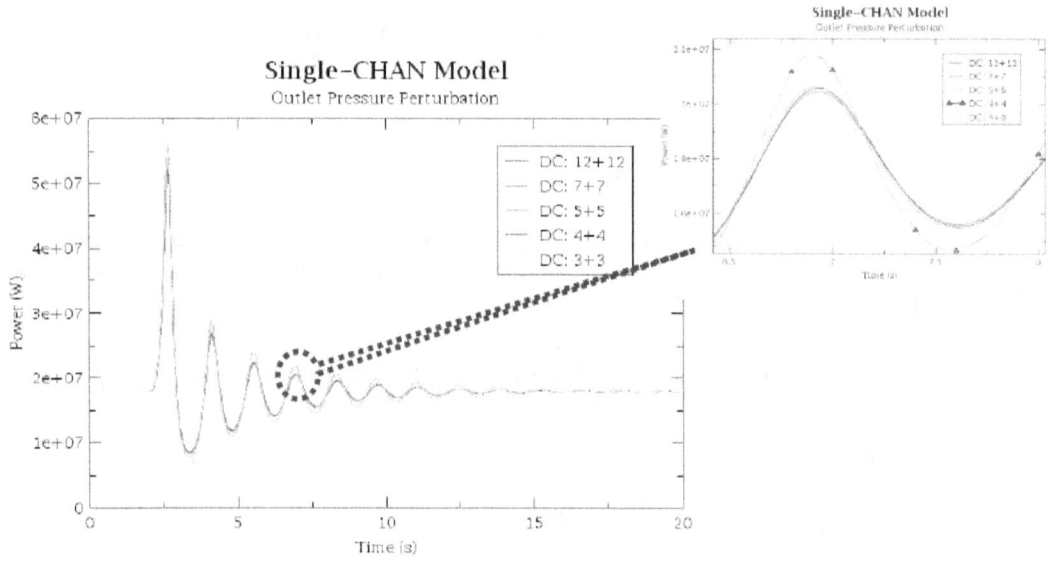

Figure 23. Response to a pressure perturbation for single-CHAN model.

4.2.2. Results for Oskarshamn Core-wide Stability

Steady-state results for each DC branch case are summarized in Table 8. Figure 24 shows steady-state core axial DC profiles. The core exit average DC is about 80%. The coolant in the top of a hot fuel channel can be up to 90% void. Figure 25 shows core flow oscillations induced by a control rod perturbation. Again, it shows clearly that the case of DC 4+4 differs significantly from the other cases when the 100% void branch is removed.

Table 8. Steady-state parameters for Oskarshamn

	Steady-state			
Parameters	DC: 12+12	DC: 7+7	DC: 5+5	DC: 4+4
Keff	0.901057	0.901236	0.90158	0.900782
Pxyz	2.138	2.132	2.12	2.144
Pxy	1.482	1.482	1.482	1.481
Pz	1.314	1.309	1.298	1.317

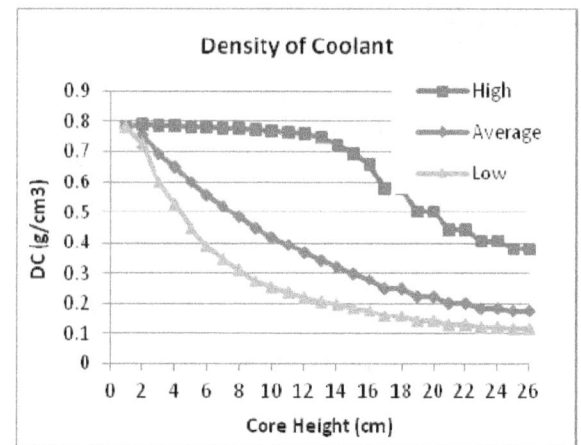

Figure 24. Core axial coolant density for Oskarshamn

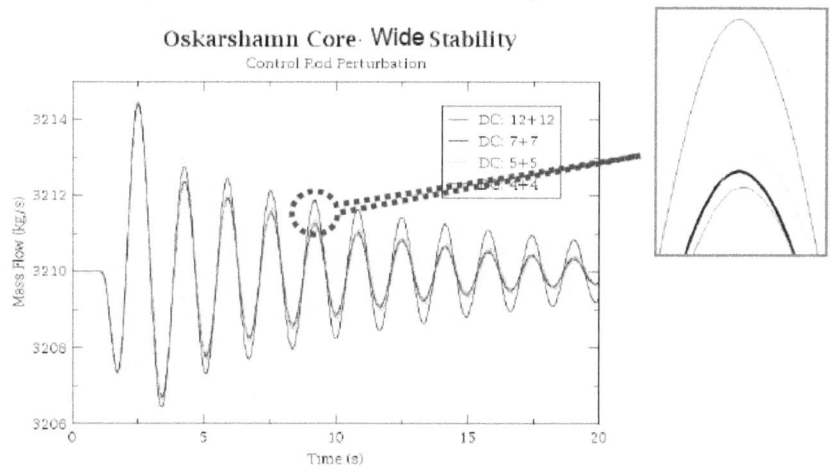

Figure 25. Response to a control rod perturbation for Oskarshamn

4.2.3. Results for Ringhals Core-wide Stability

Steady-state results for each DC branch case are summarized in Table 9. Figure 26 shows steady-state core axial DC profiles. The core exit average DC is about 75%. For this case, the coolant in the top of a hot fuel channel can be up to 85% void. Figure 27 shows core flow oscillations induced by a control rod perturbation. The case of DC 5+5 shows some deviation from the reference case and the case of DC 7+7. The case of DC 4+4 differs significantly from the other cases when the 100% void branch is removed.

Table 9. Steady-state parameters for Ringhals

	Steady-state			
Parameters	**DC: 12+12**	**DC: 7+7**	**DC: 5+5**	**DC: 4+4**
Keff	0.925328	0.92542	0.92595	0.925837
Pxyz	2.386	2.382	2.357	2.361
Pxy	1.688	1.688	1.688	1.687
Pz	1.163	1.163	1.171	1.172

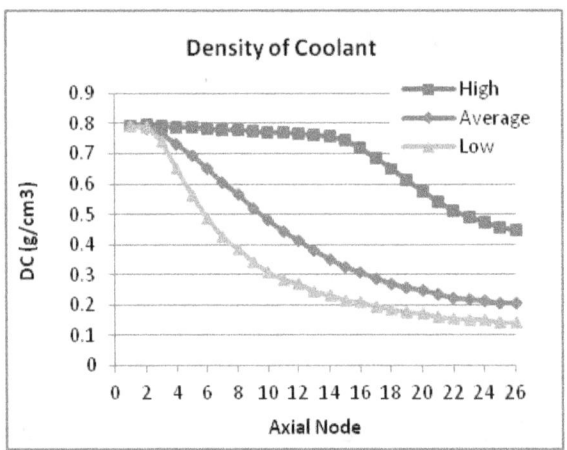

Figure 26. Core axial coolant density for Ringhals.

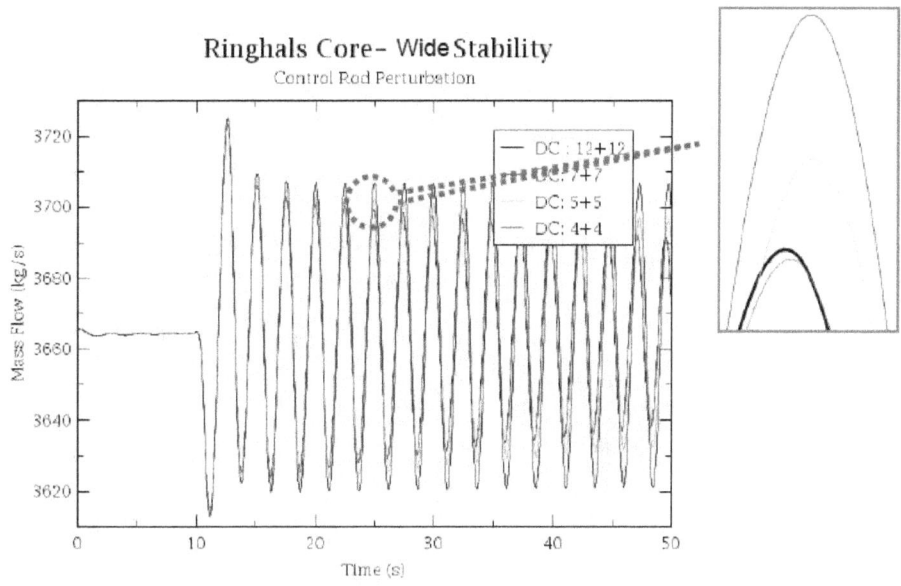

Figure 27. Response to a control rod perturbation for Ringhals.

4.2.4. 4.2.4 Conclusions of DC Branch Effects

Based on the above analysis of the DC branch effects, the following conclusions are drawn.

1. BWR steady-state variables (i.e., K_{eff}, P_{xyz}, P_{xy}, and P_z) are not highly sensitive to the DC branches. However, results of DC 7+7 (20% void spacing) have the best agreement with the reference case DC 12+12.

2. BWR transients show sensitivity to the DC branches, particularly in the core upper region where the void fraction is above the range of the DC branches.

29

5.0 EFFECTIVE DELAYED NEUTRON FRACTION

5.1. DEFINITION OF β_{eff}

The energy collapse of fine group XS data into few group constants requires proper weighting of the fine-energy and -space fluxes and reaction rates. Typically, XS are weighted by the local flux to maintain constant reaction rates. However, integral variables like the delayed neutron fraction, β_{eff}, require weighting by both the forward and adjoint fluxes, also known as bilinear weighting. The following relation provides the optimal bilinear weighting to collapse the fine-energy and -space fluxes for β_{eff}.

$$\beta_{eff,i} = \frac{\sum_m \int dr \int dE \chi_{d,i}^m(E)\phi^*(r,E) \int dE' \beta_i^m v \Sigma_f^m \phi(r,E')}{\sum_m \int dr \int dE \chi_t^m(E)\phi^*(r,E) \int dE' v \Sigma_f^m \phi(r,E')} =$$
$$\frac{\sum_m \beta_i^m \int dr \int dE \chi_{d,i}^m(E)\phi^*(r,E) \int dE' v \Sigma_f^m \phi(r,E')}{\sum_m \int dr \int dE \chi_t^m(E)\phi^*(r,E) \int dE' v \Sigma_f^m \phi(r,E')}, i = 1,..,6 \quad (5.1)$$

where

> m = the mth isotope
>
> $\phi^*(r,E)$ = adjoint flux
>
> $\phi(r,E)$ = neutron flux
>
> β_i = delayed neutron fraction of the ith energy group for fissions of isotope m
>
> $\chi_{d,i}^m(E)$ = delayed neutron spectrum of the ith energy group for fissions of isotope m
>
> $\chi_t^m(E)$ = total fission energy spectrum (prompt + delayed)
>
> $v \Sigma_f \phi(r,E')$ = total fission neutrons per fission (prompt + delayed)

If $\phi^*(r, E)$ is assumed constant, we obtain the "fission-weighted" β_{eff} as in Eq. (5.2).

$$\beta_{eff,i} = \frac{\sum_m \beta_i^m \int dr \int dE' v \Sigma_f^m \phi(r,E')}{\sum_m \int dr \int dE' v \Sigma_f^m \phi(r,E')}, i = 1,...,6 \quad (5.2)$$

5.2. CALCULATION OF β_{eff} USING A SINGLE-ASSEMBLY LATTICE

In practice, the problem is more complex because the fine-energy and -space fluxes are not known exactly. They must be approximated based on the results of single assembly lattice calculation. Lattice codes like TRITON, CASMO, or LANCER only calculate the fine-energy and -space fluxes for a single bundle, and they represent the core surrounding the bundle via boundary conditions (BCs). Two BCs are typically used:

1. **Reflective BC (denoted Infinite BC)**. The bundle is assumed to be surrounded by a large number of similar bundles, so any neutron leakage that occurs will be compensated by neutrons that leak from surrounding bundles into the specified one. The BC used for this case is a pure reflective BC, where all the neutrons leaked are

immediately returned to the bundle. This calculation produces an eigenvalue, which is appropriately labeled K_{inf}.

2. **Reflective BC with Critical Spectrum (denoted Critical BC)**. In reality, cores are typically loaded in a checkerboard pattern, in which high-reactivity bundles are surrounded by low-reactivity bundles (once- or twice-burned); therefore, net leakage occurs from the high-reactivity to the low-reactivity bundles. To compensate for this effect, the critical BC adds the concept of a material buckling, B_m^2, which acts as an absorber and effectively removes neutrons from the high-reactivity bundles and adds them to the low-reactivity ones. For low-reactivity bundles, the material buckling is negative and effectively reduces the fraction of neutrons absorbed in the bundle; for high-reactivity bundles, the material buckling is positive and increases the absorption fraction.

The critical BC physically represents more accurately the neutron balances in a real reactor, and it has been proved in practice that reaction rates calculated by the critical BC (i.e., using the material buckling approximation) are a better approximation than rates calculated with the infinite BC.

5.3.   PHYSICAL DESCRIPTION OF THE PHENOMENA

The adjoint calculation, however, may be a bit counterintuitive. The adjoint, Φ^+ (r, E), is not related to a production rate, but it is defined as the probability that a neutron of energy E at location r will produce an additional fission neutron. The following example illustrates the process. For this example, we have two bundles side by side, as shown in Figure 28.

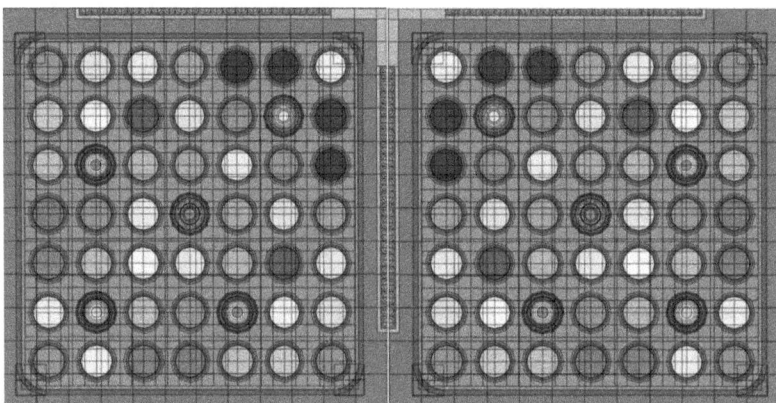

Figure 28. Two bundle sample configuration.

Figure 29 represents a hypothetical neutron path in which a neutron is born on the left bundle but is leaked and interacts in the right bundle, where it produces either a capture or a fission. If the final interaction is a fission event, this path would contribute to the adjoint flux at the path originating point; if the neutron is absorbed, it does not.

Since lattice calculations are performed on a single lattice, we must model this type of neutron path using an approximation through a BC. In all cases (either infinite or critical approximations), lattice codes use reflective BCs, as illustrated in Figure 30.

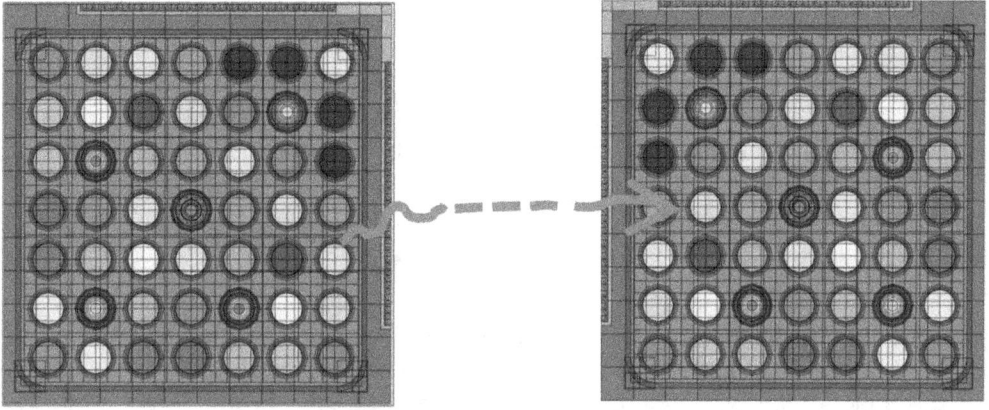

Figure 29. Hypothetical neutron path born on left bundle and absorbed on the right one.

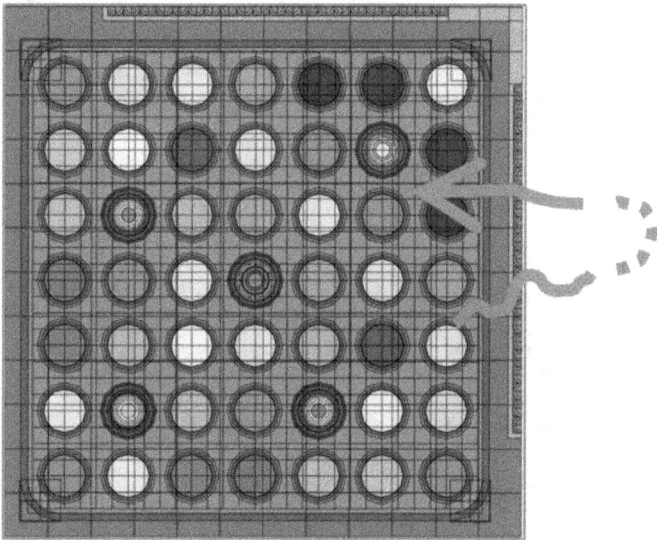

**Figure 30. Illustration of single-lattice approximation,
where the neutron path is reflected back into the bundle.**

Both approximations (Infinite and critical) simulate the neutron path of Figure 29 by reflecting it as in Figure 30. In the case of the infinite approximation, the neutron is allowed to interact with the bundle. In the case of the critical approximation, the bundle absorption is modified with a material buckling, B_m^2, that is adjusted to make the bundle critical. The question is which of the two approximations more closely reflects the real adjoint flux?

Let us assume a limiting case in which the control rod is inserted. In this case, the calculated eigenvalue for a single lattice will be significantly subcritical, and a large material buckling correction will be required for the critical approximation. Thus the control bundle scenario will result in the largest difference between the two approximations. We need to remember that the adjoint flux we are trying to estimate is the probability that at the end of the neutron path, there will be a fission-induced neutron or a capture.

If we follow the neutron path on the real reactor (Figure 29), we see that the neutron exits the left bundle through the left side of the control rod and then must cross the right half of the control rod before entering the right bundle. If the neutron is absorbed by either the left or the right side of the control rod, this event counts as a capture and the adjoint value is reduced. If the neutron makes it through the control rod, then the probability of fission in the bundle, which depends on its ratio of ^{235}U to other absorbers, need not be corrected by the presence of the control rod because the rod was taken into account explicitly when the path was calculated. These physical phenomena are more closely simulated by the infinite approximation, in which the neutron path must cross the left side of the control rod twice (one to the right and once again to the left to enter the bundle); therefore, the bundle absorption and fission cross sections are unmodified.

If we use the critical approximation, we must decrease the absorption in the bundle using a negative material buckling (because K<1 in a controlled bundle). The net effect of this negative buckling on the critical approximation is to artificially increase the probability that the neutron path will create a fission neutron, which would bias the adjoint flux in the positive direction.

Figure 32 and Figure 31 illustrates the effect of neutron leakages from other bundles. In this particular example, we assume that the top and bottom neighbors have higher reactivity and leaked neutrons into our bundle, but the left and right bundles have lower reactivity and receive a net leakage. When we ask "what is the probability the red neutron path will generate a fission?" (i.e., what is the adjoint), the answer does not change, and we see that the leakage pattern is second order and does not impact the adjoint flux.

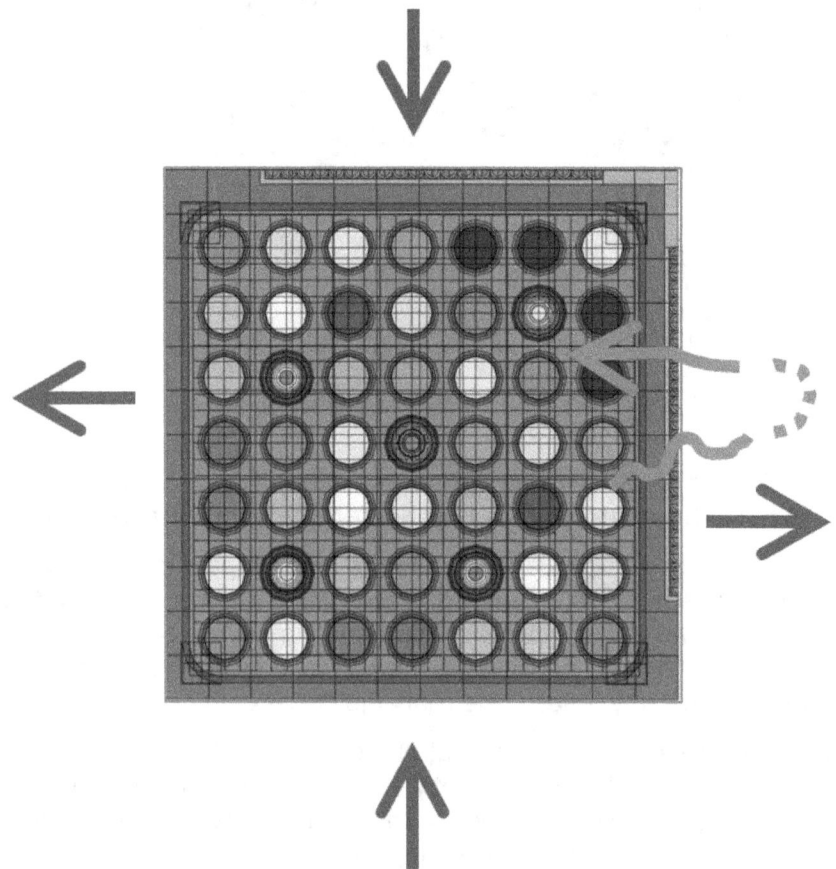

Figure 31. Illustration of the effect of neutrons leaked from surrounding bundles.

5.4. NUMERICAL SIMULATION

Calculations indicate that the β_{eff} calculated with adjoint critical flux weighting (or K-ratio weighting) could be ~20% lower for rodded assemblies than for unrodded ones. However, β_{eff} calculated with adjoint infinite flux weighting for rodded assemblies is almost the same as for unrodded ones. This difference of 20% in β_{eff} is very significant because it affects the reactivity directly when measured in dollars. For the same perturbation, if β_{eff} is 20% lower, the reactivity and associated transient power response is 20% higher.

A numerical study was performed to confirm that the infinite approximation yields a better adjoint value for β_{eff} weighting. For this study, a "mini core" in a 2×2 colorset configuration was used, as shown in Figure 32. The colorset contains two types of assemblies: Assembly A is a low-burnup assembly with high reactivity, and Assembly B is high-burnup with low reactivity; thus there is a net leak of neutrons from the A assemblies to the B assemblies. It is assumed that this colorset will repeat itself in the core for a typical checkerboard pattern. Since to the right side of the B assembly there would be an A assembly, the simulation used periodic BCs, which are more realistic for this colorset. For this calculation, all bundles are unrodded (the control rod is fully out).

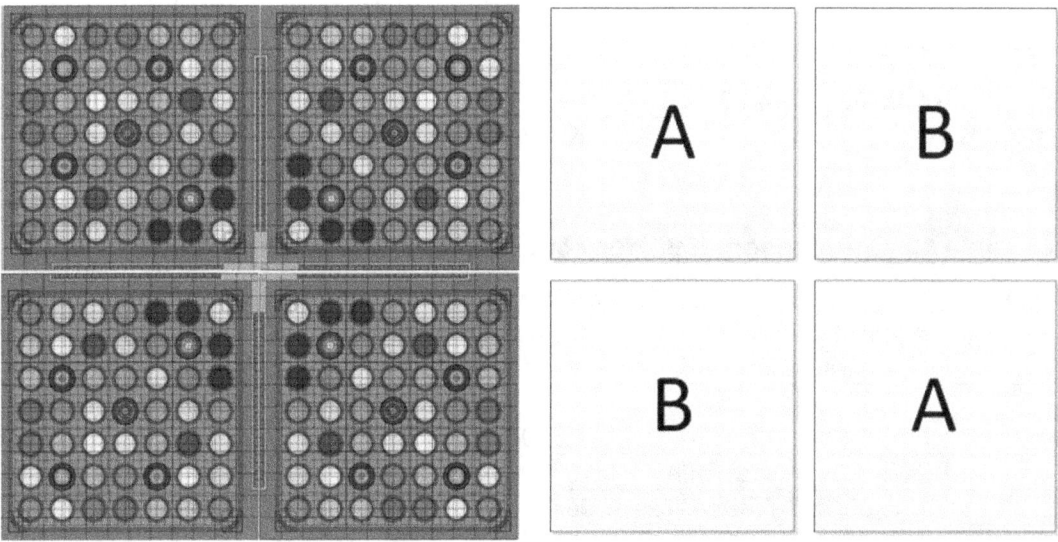

Figure 32. Example colorset configuration.

In this study, all of the assemblies in the colorset are the same Peach Bottom type-3c fuel used in previous sections. Three different burnup and void configurations were considered for the study, as shown in Table 10. Separately, single-bundle lattice calculations were performed for each of the configurations and assembly types.

The TRITON code was used to calculate β_{eff} for each assembly, using the neutron flux $\phi(r, E)$ and its adjoint flux $\phi^*(r, E)$ directly calculated in the above colorset configuration. It was compared with the β_{eff} calculated from single assembly lattice calculations using the two approximations based on the calculated adjoint critical flux and the adjoint infinite flux for the single assembly. The β_{eff} results are summarized in Table 11 through Table 13 along with the eigenvalues for each of the configurations modeled.

Table 10. Three colorset burnup configurations studied

Configuration	A Burnup (GWd/MTU)	A Coolant density (g/cc)	B Burnup (GWd/MTU)	B Coolant density (g/cc)
C1	3.79	0.73808 (0% void)	24.70	0.73808
C2	7.25	0.45721 (40% void)	31.11	0.45721
C3	8.40	0.24657 (70% void)	27.50	0.24657

Table 11. Comparison of β_{eff} for configuration C1—low burnup case

Group	Single Assembly — Adjoint critical flux weighting A	B	Single Assembly — Adjoint infinite flux weighting A	B	Colorset A	B
1	1.83E-04	1.12E-04	1.78E-04	1.17E-04	1.78E-04	1.17E-04
2	1.29E-03	9.61E-04	1.26E-03	1.00E-03	1.25E-03	1.01E-03
3	1.17E-03	8.24E-04	1.14E-03	8.60E-04	1.13E-03	8.64E-04
4	2.39E-03	1.65E-03	2.33E-03	1.72E-03	2.32E-03	1.73E-03
5	8.17E-04	6.36E-04	7.97E-04	6.63E-04	7.93E-04	6.67E-04
6	2.85E-04	2.02E-04	2.78E-04	2.11E-04	2.77E-04	2.12E-04
Total	**6.13E-03**	**4.39E-03**	**5.98E-03**	**4.58E-03**	**5.95E-03**	**4.60E-03**

Note: K_{inf} of A: 1.0590; K_{inf} of B: 0.9092; K_{inf} of colorset: 0.9854.

Table 12. Comparison of β_{eff} for configuration C2—medium burnup case

Group	Single assembly — Adjoint critical flux weighting A	B	Single assembly — Adjoint infinite flux weighting A	B	Colorset A	B
1	1.72E-04	1.04E-04	1.65E-04	1.09E-04	1.64E-04	1.10E-04
2	1.24E-03	9.29E-04	1.19E-03	9.70E-04	1.18E-03	9.79E-04
3	1.12E-03	7.99E-04	1.08E-03	8.34E-04	1.07E-03	8.41E-04
4	2.28E-03	1.61E-03	2.19E-03	1.68E-03	2.18E-03	1.70E-03
5	8.04E-04	6.47E-04	7.75E-04	6.75E-04	7.70E-04	6.80E-04
6	2.78E-04	2.03E-04	2.68E-04	2.12E-04	2.66E-04	2.13E-04
Total	**5.89E-03**	**4.29E-03**	**5.67E-03**	**4.48E-03**	**5.63E-03**	**4.52E-03**

Note: K_{inf} of A: 1.0952; K_{inf} of B: 0.9023; K_{inf} of colorset: 0.9987.

Table 13. Comparison of β_{eff} for configuration C3 – high burnup case

Group	Single assembly				Colorset	
	Adjoint critical flux weighting		Adjoint infinite flux weighting			
	A	B	A	B	A	B
1	1.58E-04	1.03E-04	1.54E-04	1.07E-04	1.53E-04	1.08E-04
2	1.17E-03	9.18E-04	1.14E-03	9.55E-04	1.13E-03	9.64E-04
3	1.06E-03	8.00E-04	1.03E-03	8.33E-04	1.02E-03	8.39E-04
4	2.16E-03	1.63E-03	2.11E-03	1.69E-03	2.10E-03	1.70E-03
5	7.98E-04	6.73E-04	7.80E-04	7.00E-04	7.76E-04	7.04E-04
6	2.72E-04	2.12E-04	2.65E-04	2.21E-04	2.64E-04	2.22E-04
Total	**5.62E-03**	**4.33E-03**	**5.49E-03**	**4.51E-03**	**5.45E-03**	**4.54E-03**

Note: K_{inf} of A: 1.0590; K_{inf} of B: 0.9092; K_{inf} of colorset: 0.9818.

The colorset calculations indicate that, overall, the adjoint infinite flux weighted β_{eff} gives more accurate results (~1% error) than the adjoint critical flux option (~5% error). Therefore, the recommendation from this study is to use the infinite approximation to estimate the adjoint flux for β_{eff} weighting.

Differences in the actual implementation of the β_{eff} weighting in lattice codes (CASMO, HELIOS, TRITON, and LANCER02) may result in measurable differences between the β_{eff} values provided by different codes for the same conditions.

6.0   CODE-TO-CODE COMPARISON

Each of the three TRACE/PARCS models was run with the CASMO, HELIOS, and TRITON XS, respectively. Steady-state results are summarized in Table 14, and transient results are shown in Figure 33, Figure 34, and Figure 35. Note that delayed neutron fractions of each XS set used for this comparison analysis are fission-weighted.

Comparisons for the steady-state variables show good agreement between the XS sets. Differences in K_{eff} are less than 200 pcm for the single-CHAN case, 150 pcm for the Oskarshamn, and 580 pcm for the Ringhals. In addition, the maximum relative difference in all power peaking factors is less than 3%.

Comparisons for the transients show good agreement between the CASMO and HELIOS XS sets. Poor agreement of TRITON is believed to be due to the over predicted fission-weighted β (because of a code bug in the fission weighting, see Figure 8).

Table 14. Steady-state parameters

Parameters	Steady-state								
	Single-CHAN			Oskarshamn			Ringhals		
	CASMO	HELIOS	TRITON	CASMO	HELIOS	TRITON	CASMO	HELIOS	TRITON
K_{eff}	0.99915	1.0014	1.00102	0.901	0.90156	0.90231	0.9253	0.931071	0.92919
P_{xyz}	1.773	1.766	1.792	2.14	2.189	2.127	2.388	2.389	2.302
P_{xy}	1	1	1	1.482	1.488	1.482	1.688	1.683	1.685
P_z	1.773	1.766	1.792	1.315	1.345	1.295	1.164	1.195	1.167

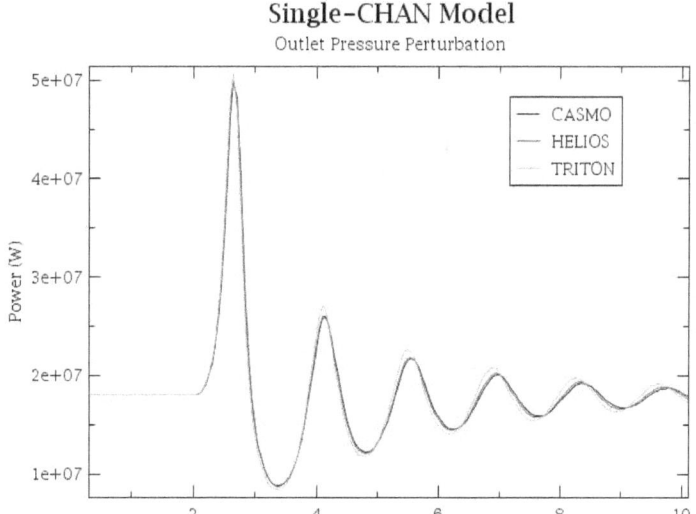

Figure 33. Single channel results.

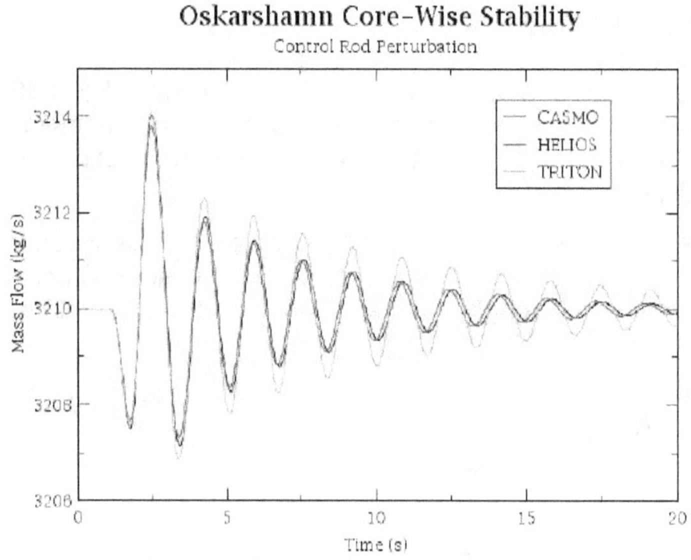

Oskarshamn Core-Wise Stability
Control Rod Perturbation

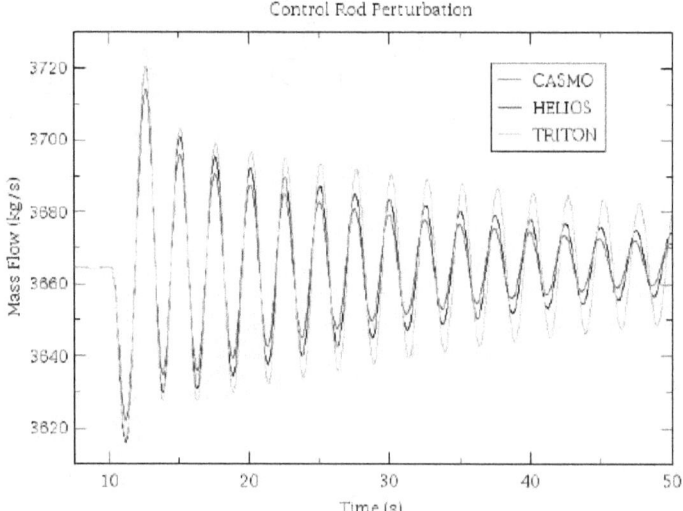

Figure 34. Oskarshamn results.

Ringhals Core-**Wide** Stability
Control Rod Perturbation

Figure 35. Ringhals results.

7.0   ISSUES WITH TWO-GROUP KINETICS EQUATIONS

There is a degree of inconsistency in the traditional formulation of two-group kinetics equations. In the traditional formulation, the two-group XS data are obtained by an energy collapse with forward flux weighting, while β is weighted with bilinear adjoint flux and forward flux weighting. This inconsistency in the traditional formulation of two-group kinetics equations is a well-known issue. One of the major problems with this formulation is that it has considered the relative importance difference of delayed neutrons but has not considered the effects of scattering terms on the neutron importance. Some researchers [6, 7] have suggested that bilinear weighting of the collapsed XS be used to provide a consistent and rigorous mathematical foundation for the formulation of the two-group kinetics equations. Unlike the linear weighting method, bilinear weighting employs both the forward and adjoint spectra for group collapsing, including 1/v, XS, and β.

Wade and Bucher [6] performed a detailed analysis of bilinear-weighted XS. They found that the errors in reactivity worth of scattering materials can be reduced significantly when XS are weighted bilinearly (adjoint flux and forward flux). Lee et al. [7] performed an assessment study of the bilinear weighted two-group kinetics equations with a 1-dimensional, fine-mesh diffusion code. Their study shows that bilinear weighing of the two-group kinetics equations has good agreement with a 97-group formulation of kinetics equation.

For the calculation of β_{eff}, our review of XS methods has discovered differences in the actual implementation of Eq. (3.4) in lattice codes (CASMO, HELIOS, TRITON, and LANCER02), which may result in measurable differences between the β_{eff} values provided by different codes for the same conditions. For example, for some configurations, the β_{eff} calculated with adjoint critical flux weighting or K-ratio weighting could be up to 20% lower for rodded assemblies than for unrodded ones. However, the β_{eff} calculated with adjoint infinite flux weighting for rodded assemblies is almost the same as for unrodded ones. The difference of up to 20% in β_{eff} can result in errors of up to 20% in transient power.

Two-group spatial kinetics equations are extensively used for light water reactor transient analysis, and are the methodology used by TRACE-PARCS. The inconsistency between the methodologies for generating β_{eff} and the traditional formulation of the two-group cross sections, may pose some concerns about the accuracy of transient analysis. However, the sensitivity analyses performed in Section 5.4 indicate that these inconsistency errors are minimized if the adjoint infinite flux weighting is used for β_{eff} generation, which is the recommended methodology used in SCALE/TRITON and CASMO. For the cases studied in in Section 5.4, this error is of the order of a few percent, but it affects mostly the controlled bundles, which contribute less to the transient power; thus we judge this error acceptable for confirmatory calculations.

8.0   BWR PARCS CROSS SECTIONS GUIDANCE

This section provides guidance on how to choose XS histories and branches for BWR analysis. The guidance herein is applicable to all BWR designs.

For BWR steady-state and transient analysis, PARCS uses two-energy-group XS for each computational node in the 3-dimensional grid. The PARCS XS are parameterized as a function of four instantaneous state variables: CR, TF, DC, and PC. The XS also depend on the isotopics (i.e., concentration of ^{235}U, ^{239}Pu ...), which are parameterized as a function of control rod and moderator density history variables.

In a typical calculation, the fuel temperature, moderator density, and soluble boron concentration are calculated by TRACE for a coupled TRACE/PARCS analysis. The instantaneous CR is provided by the user in the input deck. The historic control rod and moderator density values are provided by a steady-state core-follow simulator, which has followed the core operation since the initial loading up to the time of the transient to be calculated with TRACE/PARCS.

It is obvious that XS tabulation plays an important role in the coupled neutronics/thermal-hydraulics analysis. A good XS tabulation can reduce the error caused by XS and therefore improve the accuracy of the analysis. The XS tabulation defines the XS data set structure, that is, what histories and branches are considered and how many of them should be included in the XS set. Generally speaking, XS should be developed based on the specific transient to be analyzed. For example, for a BWR analysis, HDC and branch XS should cover the whole range of core DC change under both the steady-state and the expected transient conditions to avoid the need for table extrapolation.

In a typical BWR core, the DC changes significantly from the core inlet to the outlet. The coolant at the core inlet is subcooled with a density of ~760 kg/m^3, and the DC is reduced to ~200 kg/m^3 on average at the top of the core with a void faction of ~75%.

Therefore, for a BWR core operated under such a wide distribution of DC, history effects of DC on XS due to the neutron spectrum difference should be considered when generating XS. In addition, the effects of HCR on XS should be considered as well.

A BWR core usually consists of hundreds of fuel assemblies. A fuel assembly typically is made up of several axial segments with different enrichment levels and burnable poisons. For each fuel segment, two-group PARCS XS including histories and instantaneous branches are generated using a lattice physics code (e.g., TRITON, CASMO, or HELIOS). A processing code, GenPMAXS, is employed to process the two-group XS and kinetics data generated with the lattice code calculations and store them in PMAXS format for each fuel segment; the PMAXS files can be read by PARCS directly. An XS file in PMAXS format provides all of the data necessary to perform PARCS simulations for steady-state and transient applications. These PMAXS data consist of two-group macroscopic XS, microscopic XS of Xe/Sm, group-wise discontinuity functions (DF) for higher-order solution methods, and the kinetics data (e.g., delayed neutron parameters and inverse velocities). The lattice code process to generate XS is described in detail in Spriggs et al. [2, 8].

The structure of PMAXS for a sample BWR fuel type is given in Figure 36. This sample PMAXS file contains three HDCs at DCs corresponding to 0%, 40%, and 70% void at full pressure. Each HDC contains three instantaneous DC branches, three TF branches, and 3 soluble boron concentration branches. The PMAXS data structure shown in Figure 36 is repeated for control and uncontrolled bundles for each burnup step.

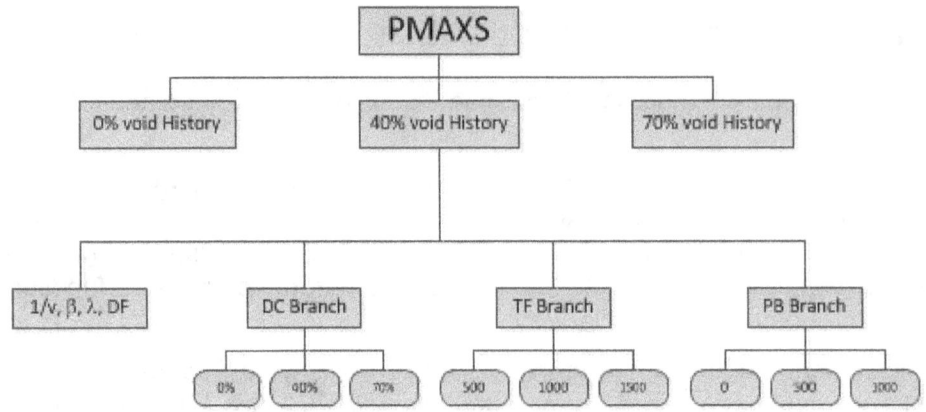

Figure 36. Typical PMAXS structure.

8.1.   HISTORY XS

Macroscopic XS have a strong dependence on history effects of DC and control rod position. Figure 37 shows Σ_a^1 and $\nu\Sigma_f^1$ of a BWR assembly as a function of history void faction at BOL and EOL, respectively. The black dashed line on each figure is the second-order polynomial fitting of the XS data at 0, 40, and 70% void, and the blue solid line is the XS calculated by the lattice code at that void.

It is seen that both Σ_a^1 and $\nu\Sigma_f^1$ curves show more nonlinearity at high void fraction (>70%) than at low void fraction, so the extrapolation of the XS above the 70% void may overestimate the XS at high void faction (see Figure 36). However, a recent sensitivity study of HDC effects concludes that neither BWR steady-state nor transient analysis is highly sensitive to HDC and HCR effects. It is found that three HDCs at 0, 40, 70% (or 80%) voids are sufficiently accurate for most BWR applications. A history point at the 90% void would be recommended for high-power density reactors to improve the XS interpolation and/or extrapolation in the core upper region with a void fraction of above 80%.

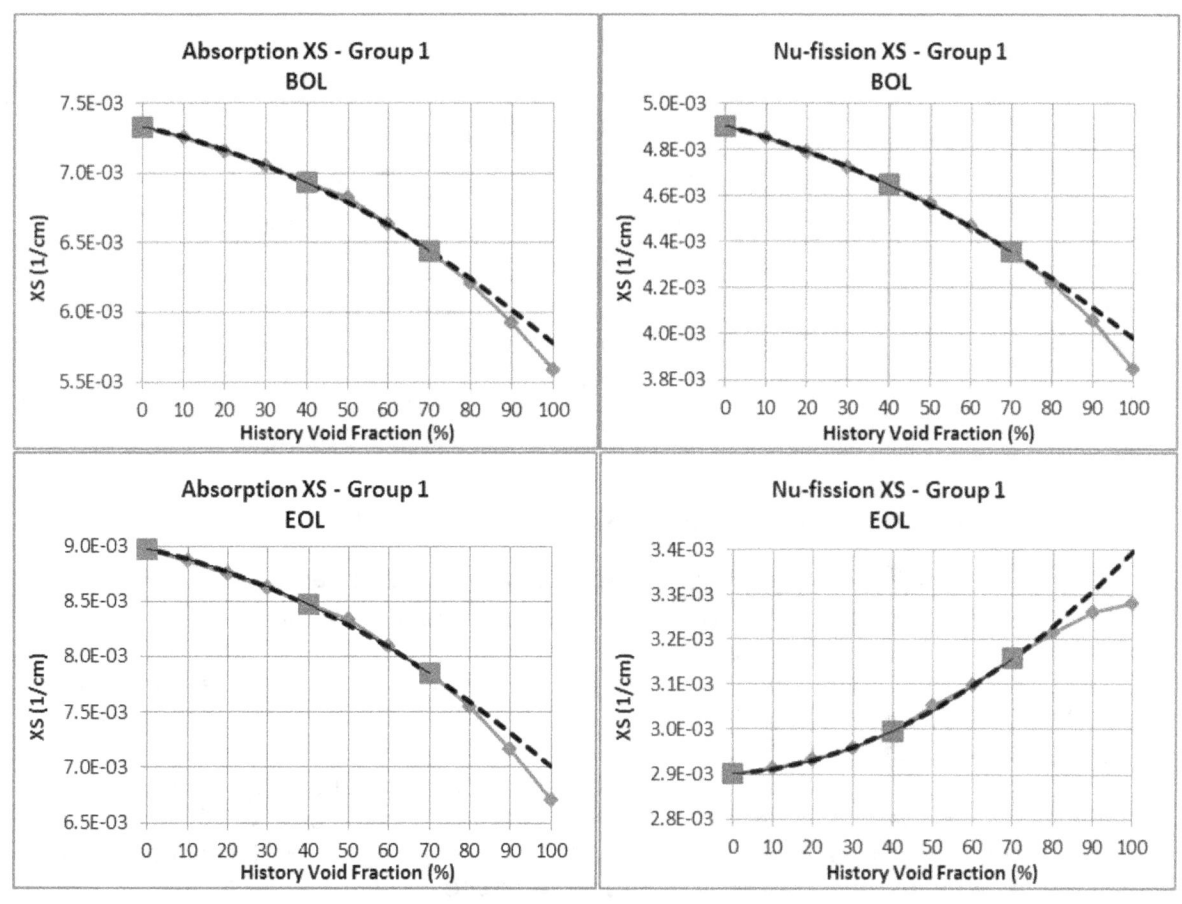

Figure 37. Void history effect on Σ_a^1 and $\nu\Sigma_f^1$.

8.2.   BRANCH XS

At each time-step of a transient calculation, XS for each PARCS neutronic node are updated based on the instantaneous value of DC, CR, TF, and boron concentration at that node. The instantaneous XS are obtained by interpolation/extrapolation between branch XS tabulated in the PMAXS files. Figure 38 shows Σ_a^1 and $\nu\Sigma_f^1$ of a BWR assembly as a function of branch void at EOL [1]. The BOL Σ_a^1 and $\nu\Sigma_f^1$ are the same as those shown in Figure 2. The black dashed line on each figure is the second-order polynomial fitting of the XS at 0, 40, and 70% void.

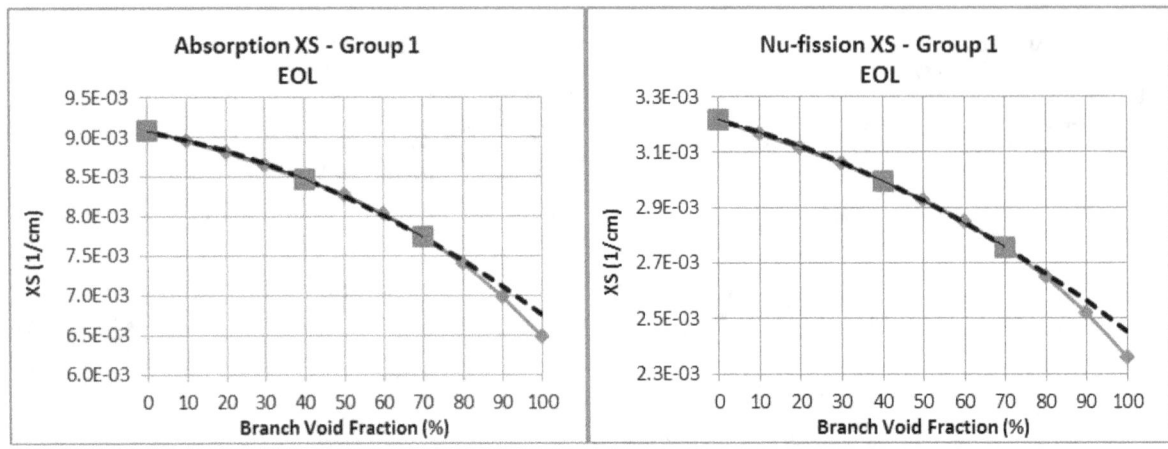

Figure 38. Brach void fraction effect on Σ_a^1 and $\nu\Sigma_f^1$.

At high void fraction (>70%) both Σ_a^1 and $\nu\Sigma_f^1$ curves show more nonlinearity than at low void fraction, so the extrapolation of the XS from the 70% void may overestimate the XS at high void fraction; however sensitivity studies show that BWR steady-state analysis is not very sensitive to the DC branch effects [1]. The same sensitivity analysis shows that transient results may be sensitive to the DC branches, especially when the DC of the core upper region is significantly larger than that of the upper branch void. The study shows that four DC branches at 0, 40, 70, and 90% voids are recommended for high-power-density BWR transient analysis. Most current-generation lattice codes cannot model the 100% void case accurately because of its extreme hard spectrum, but this result is dependent on the lattice methodology used. It is suggested that the same DC branches be used for the controlled DC branches.

For TF branches, three branches at 500, 950, and 1500 K are sufficient to calculate the TF reactivity feedback. In addition, a DC branch with DC = 1000 kg/m^3 is needed to model shutdown conditions accurately. If boron injection is modeled in a BWR transient analysis, XS boron branches should be included.

8.3.   GUIDELINES

For most BWR applications, the XS branch structure of Table 15 and Table 16 is recommended; however, specific studies may require a different structure. The user must review the requirements and ensure the applicability of the XS set to the particular application. For anticipated-transient-without-scram calculations, an additional set of boron concentration branches are needed, but that topic was not part of this study.

For β_{eff} calculations, it is recommended that the infinite adjoint approximation be used to weight the fine-energy and -space results.

Table 15. Recommended history branches

History coolant density (kg/m^3)	Percentage void at 7 Mpa (%)	Control rod	Fuel temperature (K)
738	0	0	950
457	40		
247	70		
738	0	1	
457	40		
247	70		

Table 16. Recommended instantaneous branches

Instantaneous coolant density (kg/m^3)	Percentage void at 7 Mpa (%)	Control rod	Fuel temperature (K)
738	0	0	950
457	40		
247	70		
106	90		
738	0	1	
457	40		
247	70		
106	90		
738	0	0	500
457	40		
247	70		
106	90		
738	0	1	
457	40		
247	70		
106	90		
738	0	0	1500
457	40		
247	70		
106	90		
738	0	1	
457	40		
247	70		
106	90		
1000		0	950
			500
			293
		1	950
			500
			293

9.0 REFERENCES

1. *Core Design and Operation Data for Cycles 1 and 2 of Peach Bottom 2*, EPRI NP-563, Electric Power Research Institute, June 1978.

2. G. D. Spriggs et al., "Calculation of the delayed neutron effectiveness factor using ratios of K-eigenvalues," *Annals of Nuclear Energy* **28** (2001) 477–487.

3. TRITON source code, SCALE 6.1.

4. CASMO-4 User's Manual, SSP-01/400 Rev 3.

5. Y. Xu and T. Downar, *GenPMAXS— Code for Generating the PARCS Cross Section Interface File PMAXS*, PU/NE-00-20, Purdue University, April 2009.

6. D. C. Wade and R. G. Bucher, "Conservation of the adjoint neutron spectrum by use of bilinear-weighted cross sections and its effect on fast reactor calculations," *Nuclear Science and Engineering* **64**, 517–538 (1977).

7. C. Lee, T. J. Downar, K. O. Ott, "An assessment of consistent bilinear weighted two-group spatial kinetics for MOX fuel applications," in *PHYSOR 2000 : 2000 ANS International Topical Meeting on Advances in Reactor Physics and Mathematics and Computation into the Next Millennium* (2000).

8. B. Ade, *SCALE/TRITON Primer: A Primer for Light Water Reactor Lattice Physics Calculations*, NUREG/CR-7041/ ORNL/TM-2011/21, November 2012.

NRC FORM 335 (12-2010) NRCMD 3.7	U.S. NUCLEAR REGULATORY COMMISSION	1. REPORT NUMBER (Assigned by NRC, Add Vol., Supp., Rev., and Addendum Numbers, if any.)
	BIBLIOGRAPHIC DATA SHEET *(See instructions on the reverse)*	NUREG/CR-7164

2. TITLE AND SUBTITLE		3. DATE REPORT PUBLISHED	
Cross Section Generation Guidelines for TRACE-PARCS		MONTH	YEAR
		June	2013
		4. FIN OR GRANT NUMBER	

5. AUTHOR(S)	6. TYPE OF REPORT
Dean Wang (ORNL) Brian J. Ade (ORNL) Andrew M. Ward (UM)	
	7. PERIOD COVERED (Inclusive Dates)

8. PERFORMING ORGANIZATION - NAME AND ADDRESS (If NRC, provide Division, Office or Region, U. S. Nuclear Regulatory Commission, and mailing address; if contractor, provide name and mailing address.)

Jose March-Leuba, ORNL
One Bethel Valley Road
P.O. Box 2008, MS-6010
Oak Ridge, TN 37831-6010

9. SPONSORING ORGANIZATION - NAME AND ADDRESS (If NRC, type "Same as above", if contractor, provide NRC Division, Office or Region, U. S. Nuclear Regulatory Commission, and mailing address.)

Division of System Analysis
Office of Nuclear Regulatory Research
U. S. Nuclear Regulatory Commission
Washington, DC 20555-0001

10. SUPPLEMENTARY NOTES
Carl Thurston, NRC project mgr

11. ABSTRACT (200 words or less)

This report documents a comprehensive comparison of cross sections calculated using different methodologies and codes, including CASMO, HELIOS, and TRITON XS. The conclusion is a guidance document on how to choose cross section histories and branches for boiling water reactor (BWR) analysis, and the methodology to collapse the fine-energy and -space fluxes calculated by the detailed lattice calculation. The guidance herein is applicable to all BWR designs.

For BWR steady-state and transient analysis, the PARCS code uses two-energy-group cross sections for each computational node in the 3-dimensional grid. The PARCS cross sections are tabulated as a function of four instantaneous state variables: (1) control rod insertion, (2) fuel temperature, (3) coolant density, and (4) soluble poison concentration. The cross section values also depend on the isotopic mixture, which is characterized as a function of control rod and moderator density history variables.

The recommendation includes the use of four instantaneous moderator density values at 0, 40, 70, and 90% void fraction at three different fuel temperatures of 500, 950, and 1500 K. For the history effect, three moderator density values at 0, 40, and 70% at a single 950 K fuel temperature provide sufficiently accurate results. For all cases, the user must generate these branches for controlled and uncontrolled bundles. In addition, a coolant density branch with a moderator density of 1000 kg/m3 is needed to accurately model boron injection.

12. KEY WORDS/DESCRIPTORS (List words or phrases that will assist researchers in locating the report.)	13. AVAILABILITY STATEMENT
BWR Stability	unlimited
BWR Cross Sections	14. SECURITY CLASSIFICATION
Lattice Physics	(This Page)
BWR Core Performance	unclassified
GenPMAXS Generation of the Purdue Macroscopic XS Set	(This Report)
TRACE TRAC/RELAP Advanced Computational Engine	unclassified
PARCS Purdue Advanced Reactor Core Simulator	15. NUMBER OF PAGES
	16. PRICE

NRC FORM 335 (12-2010)

UNITED STATES
NUCLEAR REGULATORY COMMISSION
WASHINGTON, DC 20555-0001

OFFICIAL BUSINESS

NUREG/CR-7164

Cross Section Generation Guidelines for TRACE-PARCS

June 2013